ES&T Presents

AUDIO

Troubleshooting and Repair

ES&T Presents

AUDIO

Troubleshooting and Repair

PROMPT© Publications is an imprint of Howard W. Sams & Company, A Bell Atlantic Company, 2647 Waterfront Parkway, E. Dr., Indianapolis, IN 46214-2041.

International Standard Book Number: 0-7906-1182-1
Library of Congress Catalog Number: 98-068717

Acquisitions Editor: Loretta Yates
Editors: J.B. Hall, Nils Conrad Persson
Contributing Editors: Homer L. Davidson, ES&T Staff, Jurgen Ewert, Sheldon Fingerman, Sam Goldwasser, John S. Hanson, Ron C. Johnson, Matt J. McCullar, Vaughn D. Martin, Marcel R. Railland, Lamar Ritchie, John A. Ross, Dale C. Shackelford, John Shepler, and Sam Wilson, CET
Assistant Editor: Pat Brady
Typesetting: J.B. Hall
Indexing: J.B. Hall
Cover Design: Christy Pierce
Graphics Conversion: Terry Varvel
Illustrations & Text: Supplied by *Electronic Servicing & Technology (ES&T)* Magazine, CQ Communications, Inc., 76 N. Broadway, Hicksville, NY 11801.

PRINTED IN THE UNITED STATES OF AMERICA

9 8 7 6 5 4 3 2 1

Table of Contents

Chapter 24

Troubleshooting Audio Circuits By the Numbers

By Homer Davidson 205

Chapter 25

Compact Disc Interactive (CD-I)—Part 1

By Marcel R. Rialland 217

Chapter 26

Compact Disc Interactive—Part 2

By Marcel R. Rialland 229

Chapter 1

Audio Power Amplifier Repair Practice

By Jurgen Ewert

In some cases, it is possible to repair audio power amplifiers without the schematic diagram, because of the similarity of circuits that are used and the common problems that occur. To service an amplifier without a diagram, start by locating the power transistors or the output pins of the power IC and you can start troubleshooting the power stage.

Ask your customer if he changed the wiring to the speakers before the stereo amp quit working. Very often the unprotected speaker outputs get shorted. Make sure to test the amplifier with your own speakers to eliminate the speakers as the source for trouble. If there is no sound and no noise when the amp is connected to known good speakers, there is a good possibility that the power stage is bad. Noise or a distorted signal at the speaker output might be caused by a problem in the preamplifier or driver.

Power Amp Repair Without a Schematic

One of my customers brought two identical Yamaha receivers, model CR-600, into my service center and wanted at least one of them back in operating condition. These beautiful 1960s style receivers had different problems. One of them did not work at all. I traced the problem to a bad component in the power supply. It worked after I fixed the problem in the power supply.

 The other amplifer put out very scratchy sound and not much power at the left speaker. My first bet was a bad power stage. Of course I did not have a schematic for these units and I did not bother to try to get one because of the age of the receivers.

Checking the dc voltages at the power transistors, I found that the voltage at the center node was almost 0V. With an ohmmeter I checked the collector-emitter resistance of the power transistors, 2SD371, in circuit and found that one had a C-E short. The 2SD371 was not available anymore, so I replaced both power transistors in the left channel with 2SD555s.

After I replaced the transistors I checked the isolation to the heat sink. Sometimes it happens that the isolation pad is not aligned properly, causing a short between collector and ground. I also made sure that the heat sink compound was thoroughly distributed to assure a low thermal resistance between transistor and heat sink.

Before powering the repaired unit up I checked the resistances from the transistor connections to ground comparing the values to the good channel. Although these resistance values do not mean much, you will detect differences between L and R channel, and depending on the readings you can decide if you want to power it up or not. The values looked close enough, so I powered the unit up slowly, connecting it through my variable power transformer.

The amplifier worked fine and the dc voltages at the left channel were the same as the values that I read at the right channel. Finally I checked the quiescent current, the current the power stage draws when no signal is applied. If this current is too high the power transistors could overheat. If the quiescent current is too low the THD value at low power output will be high. The value of 40mA that I read in this case was acceptable. This repair was a typical example for a number of audio power amps which I have seen.

A Repair Of An Integrated Power Amplifier

A Sherwood receiver S-2770RCP was completely dead. The fuse in the power supply was blown. Replacing the fuse did not cure the problem. When I powered the unit up slowly using the variable line transformer, the line current went up very quickly. I disconnected the power amps from the power supply. When I applied power again, the power supply operated correctly (*Figure 1-1*).

*Figure 1-1. The power amplifier in the
Sherwood S-2770RCP receiver is based on the
STK4040 IC.*

To determine if one or both of the channels of the power amp had problems, I
connected one channel at a time to the power supply. The amplifer operated
properly with the right channel connected, but after I connected the left
channel and applied power to the current was excessive at very low line
voltage.

Fortunately, I had a schematic for this unit. The portion of interest for this
service procedure is shown in *Figure 1-2*. The power stage of this amplifier
is based on the STK4040 power IC.

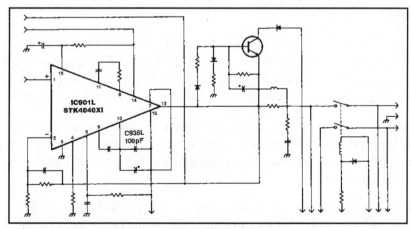

*Figure 1-2. This is a portion of the Sherwood S-2770RCP receiver that is of interest
for this service procedure. I started troubleshooting at the output (pin 13) of the
STK4040.*

To find out where the trouble was located I started at the output (pin 13) of the STK4040, measuring the resistance between pin 13 and the power voltages (pins 5 and 3) in circuit. These pins were shorted. After testing a few surrounding components I removed the IC to find out if the short was located in the STK4040 IC, or in the surrounding circuitry. The short was in the IC (*Figure 1-3*).

Figure 1-3. The problem in the Sherwood S-2770RCP receiver was a shorted audio power IC, STK4040.

Power ICs for audio amplifiers are usually expensive compared to discrete power transistors. To estimate the cost of the repair I called for the price of the IC. The first supplier quoted me a price of almost $25.00; too high for a profitable repair. I called around to see if I could find this IC at a better price and found another supplier who sold it for a little more than half of that, so I ordered one from him.

Replacing an IC in this amplifier is a little tricky because the pins do not want to line up. I found out that it is easier to insert the IC temporarily on the solder side to straighten out the pins first. That makes the job of inserting the new IC a lot easier. The receiver worked after I replaced the IC.

Because it was not clear why the IC had gone bad I checked the dc voltages comparing left and right channels. There were no differences. Finally I tested the amplifier with a sine wave input signal to watch the output signal on the oscilloscope screen.

As I watched the output signal, occasionally I saw erratic high frequency oscillations. These oscillations can cause overheating of the power stage. To prevent these oscillations, circuit designers use small capacitors in feedback paths (e.g. C936 in *Figure 1-2*). Replacement of these capacitors solved the problem.

Power Amp Shuts Off At Low Power Output

A Fisher receiver model RS-280 shut off at very low output power. My first thought was "Another bad power amplifier IC." Unfortunately, the solution was not that simple. For this receiver I did not have the schematic.

The power amplifiers of the RS-280 are ICs, STK0100 II. First I checked the power ICs for overheating but they were cool to the touch. Comparing the dc voltages at the IC pins I was not able to find any differences between left and right channels. To proceed I had to order the schematic (*Figure 1-4*).

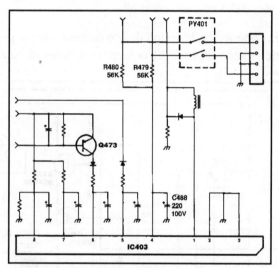

Figure 1-4. *The Fisher RS-280 has an advanced protection circuit built in (IC403). If there is something wrong in the power amps the receiver shuts down through relay PY401.*

This receiver has an advanced protection circuit built in (IC403). If there is something wrong in the power amps the receiver shuts down through relay PY401. Watching the voltages at the pins of IC403 I found that the voltage at pin 8 decreased from 5V to about 1V when I increased the volume. This pin seems to monitor the dc voltage at both power outputs through R479 and R480.

C488 (220uF/10V) is a smoothing capacitor to block audio frequencies from pin 4 of IC403. Testing C488, I found out that its capacitance was only 1.5μF instead of 220μF. After I replaced C488 with a capacitor of the correct value, the receiver put out a lot more power without shutting down. With a big load resistor on the output I tested the amplifier for the maximum power output. It continued to operate with no further problem.

An Intermittent Shutdown Problem

A Tandberg power amplifier Model TPA3006A began shutting down intermittently. Over a period of time the problem got worse. By the time the unit was brought to me, it did not turn on at all.

This amplifier is completely dc coupled. A watchdog circuit protects the speakers from dc voltages in case the power stage is out of balance.

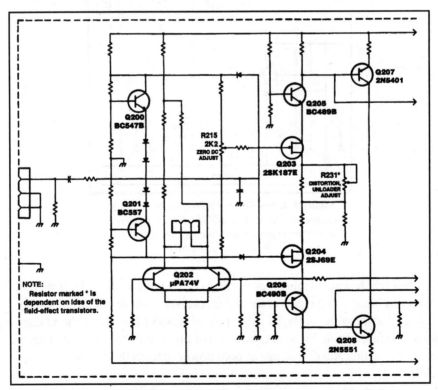

Figure 1-5. The problem in a Tandberg TPA3006A was caused by faulty transistors Q203 and Q204 in the input section.

After I removed the power amplifier board at the right channel, the left channel turned on. In the input section of each channel is a potentiometer (R215) to set the output to 0Vdc. Because of the history of this case I assumed that the amplifier became unstable and a dc voltage at the output was causing the shutdown.

My first approach was to vary the potentiometer, R215. The amplifier did turn on for a moment but there was a hissing noise in the right speaker. To make the amplifier stay on permanently I had to adjust potentiometer R231. With potentiometer R231 it is possible to adjust distortion to a minimum.

Because of the hiss, and the fact that adjusting R231 stabilized the amplifier, I suspected that the problem was caused by a component close to the input. A simple way to find a noisy transistor is spraying it with cooling spray. The noise changed when I sprayed transistors Q203 and Q204, which identified them as the cause of the problem.

It was a little tricky to replace these FETs because the manufacturer of the amplifier specifies certain Idss (drain current) groups for Q203 and Q204. I found a matching pair of FETs. I replaced Q203 (2N5458, R227=680Ω) with a 2N5459, Idss=6.4mA, R227=200Ω and Q204 (2N5461, R224=270Ω) with a 2N5462, Idss=6.3mA, R224=100Ω. After I adjusted zero dc and distortion the amplifier worked to its specifications.

By following similar logic, you may be able to solve the next problem with audiophile equipment that comes your way to be serviced.

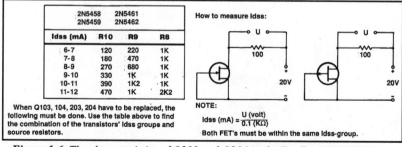

Figure 1-6. The characteristics of Q203 and Q204 in the Tandberg TPA3006A are critical. Replacements for these transistors must be carefully chosen.

Chapter 2

Better Audio Through Digital Compression

By the ES&T Staff

The world of audio has gone digital. It doesn't take a genius to see the truth of that statement. Digital compact discs (CDs) have just about elbowed vinyl LPs from the marketplace. Digital audio tape (DAT) and digital compact cassette (DCC) are here. All of this digital audio is getting listeners used to hearing the best in audio.

Unfortunately, while the sound quality of broadcasting has improved considerably over the past several years, it has not been able to keep up with the quality of recorded sound. Digital compression may soon change all that.

Digital consumer, computer and telecommunications products all have in common the need to process vast amounts of digital data at very fast rates— fast enough to appear instantaneous to users. Very high-speed digital signal processors (DSPs) are capable of processing the data load, but there are practical limits to the amount of data that can be easily stored or transmitted. Hence the need for digital compression technology.

In simple terms, digital compression refers to various software and integrated circuit hardware techniques used to "squeeze" information by removing unnecessary data. This information is then encoded and either transmitted or stored on tape or disc. The user's equipment later decodes the information and fills in the data that was removed for compression. This latter process is called "decompression," though both the compression and decompression processes are usually referred to together as "compression."

Many people associate compression with images, both still and moving. However, audio information is also compressed for digital transmission and storage. In fact, some potential applications, such as digital radio require audio compression only. With other applications, such as HDTV, audio compression opens more room, or bandwidth, for extra video information or audio channels.

These compression techniques are based on the International Standards Organization's (ISO) Joint Photographic Experts Group (JPEG) standard for still images, and Motion Picture Experts Group (MPEG) standard for audio and full-motion video.

Raw, uncompressed digital audio and video data requires too much processing to make electronic equipment cost-effective and practical for many applications. The JPEG algorithm compresses each frame of video data by removing information to which the human eye is not sensitive. MPEG extends this compression technique by transmitting only the part of each video frame image that differs from the one before it.

With JPEG technology, video can be compressed enough to store and manipulate video on a personal computer. With MPEG technology, over 70 minutes of audio and video data can be stored and played from an 80-cent compact disc, providing a less expensive distribution medium than videotape.

Products Applying Digital CompressionTechnology

Consumer electronics products that are candidates for application of digital compression technology include digital radio receivers, digital cable and direct broadcast (DBS) television decoders, HDTV, interactive compact discs (CDI), digital camcorders and videocassette recorders, digital compact cassettes and minidisks, and CD-ROM-based karaoke and video games. To be competitive in their markets, these products must be produced in high volume at low cost and have low power requirements to prolong battery life between charges.

Applications for digital compression technology also exist in the areas of personal computers and telecommunications.

Digital Compression is a Reality

It won't be long before consumer electronics service centers begin to see products with digital audio compression circuitry in them. In May, Texas Instruments introduced a new IC that will be able to produce CD quality sound in emerging consumer electronics products and entertainment systems. The device, the TMS320AV110, for use in products ranging from portable CD players to advanced digital television services, produces quality audio by decoding digital data that has been stored in a compressed format.

And while this technology can no doubt be expected to be as reliable as the rest of today's ultra-reliable consumer electronics componentry, it will fail on occasion, which will introduce new failure modes for consumer electronics servicing technicians to contend with.

An awareness of this new technology will help technicians be ready for products that employ it when it arrives on the service bench.

Chapter 3

CD Alignments
By Sheldon Fingerman

Maybe you are already repairing CD players, or just contemplating whether to get into it. As with servicing any product, one of your prime concerns is reducing callbacks. One way to minimize callbacks is to follow the manufacturer's alignment procedures. Over the long haul, the money saved by a service center in reduced callbacks could very well cover the cost of any extra equipment needed.

Another reason for checking the alignment of a CD player as part of a service routine is that specs seem to drift over time. You will find that proper alignment will not only solve some annoying problems, but will usually shorten disc access time as well.

It's really amazing how much faster a CD player can jump from track to track after proper alignment. Since deterioration of disc access time occurs over a long period of time, the customer will not only be happy that their player is fixed, but will wonder what you did to "hot-rod" it.

CD Alignment Equipment

The pieces of equipment needed to align most players are a dual-trace scope, an audio generator, a frequency counter, and a specific test disc. Although most manufacturers call for a scope of 100MHz or more, around 50MHz or 60MHz seems to work fine. The frequency counter must read at least 50MHz, and the test disc may have to be purchased from the manufacturer. Some luxury items that are nice to have are a laser power meter and a "torture" disc.

If you're contemplating working on CD players, and you do not have a frequency counter or an audio generator, one of the newer generators with a built-in counter (for use internally or externally) may be just what you are looking for.

648218

Test Discs

Some test discs are universally used among several manufacturers, with alignment specs given for the different discs. A test disc from another manufacturer will usually work just fine, but it is strongly recommended that you use a test disc from the manufacturer of the product you're aligning.

Keep in mind, however, that it should never be necessary to turn a pot very much for any adjustment. If you're using another manufacturer's test disc and you find yourself having to turn adjustment pots excessively, the problem may be the wrong test disc, or a symptom of a problem relating to that particular alignment. If you are using the wrong disc you will never know which it is.

It should be noted that in communicating with technical support, my experience has been that many of the support technicians didn't seem overly concerned that I was using another manufacturer's test disc. Most of these discs contain a variety of music (no special test tones that I can discern), and a lot of tracks, and run the information right out to the edge of the disc. You'll have to draw your own conclusions from your own experience.

Service Literature

When working on CD players, a service manual is really more of a necessity than a luxury. With the price of some service manuals on the high side, especially for non-authorized service centers, you may be tempted to wing it. If all of the adjustment points and test points are well labeled, it may be worth a try. You should, however, mark all of the adjustment pots before you touch any of them so that you can return them to their original positions— just in case.

Remember, the player should function better, not worse, when you are through.

Many of the newer carousel (turntable) type CD players require removal of the entire loading drawer to gain access to the adjustment points. This usually takes only a few minutes with the proper instructions.

As you are probably aware from servicing other devices, many different models by the same manufacturer are very similar, and a manual from one may be transposed to another. CD players are no different, so you may be able to spread the cost of one manual out over several other models.

Special tools fall into the same category. If you're pretty sure that this is the only CD player of this type, which would be easier to service with a special tool, that you will be servicing, you might be able to get by with the tools you have. But if you know that you can spread the cost of a special tool over several repairs, and it will make your servicing more efficient, it may be worth the expenditure.

Some CD Player Servicing Precautions

Before beginning you should take some precautions into account. The service manual will warn you against looking at the laser beam, and tell you not to put the laser diode in your mouth. I can understand why it might be necessary to warn you that you should not put your eyeball up against the laser lens (someone who isn't aware of the danger to eyesight might be tempted to look at it to see if they can tell if it's working), but why anyone would feel compelled to eat the laser diode is beyond me. Other, and more realistic concerns are warnings about proper handling of the laser assembly, and antistatic precautions.

Become Familiar With the Procedures

Read all of the alignment procedures before you start. This will give you a chance to see if you need any special tools or if you have to make any "filters," and let you explore the circuit board for all of the test and adjustment points (*Figure 3-1*). Some of them may be labeled so clearly that you hardly need the manual at all; others may be extremely obscure (*Figure 3-2*). It's easy to figure that a pot labeled TG is tracking gain, and a test point labeled GND is ground. Labels like VR102 and TP1 aren't going to get you very far without the proper manual.

Figure 3-1. *An example of connector type test points on the Denon DCM 777: all grouped together, easy to get to, and well labeled.*

Figure 3-2. *An example of well labeled adjustment pots. Focus offset (FO), tracking offset (TO), focus gain (FG), and tracking gain (TG) can be easily spotted on this circuit board.*

While we're on the subject of manuals, we have all seen more than one error in service literature. Most manuals are translated into English from another language. From some of the errors I've encountered, I've always felt that the same person who did the translation also proof read it. If you try a procedure that doesn't work, and you've double checked everything, the problem may be that there is an error in the manual.

A couple of methods seem to work for me when this happens. First, read ahead. Maybe you're supposed to be on test point 3, even though the manual says 2. Later, in that same procedure the manual may read, ". . . and be sure to remove the jumper from test point 3..." This would tell you that maybe 3 was the proper test point. In most such cases that I've run into, this cleared up the situation.

Second, if all the test points are lined up, resembling a connector, and numbered like a connector, they are usually labeled as well. It is not uncommon to find that the pin numbers are labeled in an order that's the reverse of the order printed in the manual's instructions.

If the instructions tell you to connect the ground of your scope to pin 5 (GND) of a 5 pin connector, but pin 5 is labeled TEO, check pin 1. If it is labeled GND, either the labels are wrong, or the manual is wrong. Use an ohmmeter to determine which pin is actually ground, and using that as a reference continue on. When in doubt, I've found that the labels on the circuit board are usually more correct than the pin numbers that are called out in the manual.

Labels can also help you when making connections. For adjusting Focus Gain you will have to both read, and inject a signal. You will be adjusting control FG, and attaching your probes to FEI and FEO. When tracing connections remember the I in FEI stand for Input, and the O in FEO stands for output. Obviously, you would not inject a signal into an output.

Using a Filter

Most CD players require some kind of "filter" for proper adjustment, and you will have to make one. They are usually composed of a resistor and capacitor, or just a resistor. If another manufacturer calls for values different from those for a filter you have already made, the one you have will probably work fine.

The reason for the filter is to reduce the amount of "fuzz" in the waveforms. Try viewing some of the waveforms with and without the filter. You will quickly see why it is such a necessity.

Aligning the Denon DCM 777

This article will use a Denon DCM 777 as an example, with references to general alignments and problems found on other players. The Denon DCM 777 is a cartridge type multidisc player, sharing many mechanical components with Pioneer models of the same type. The service manual is easy to understand, and the alignment procedures are well documented.

Before beginning alignments you will have to place the player in the test/ service mode. On this Denon you enter the service mode by shorting two pins together when powering up the player. Disc number 1 will appear in the display (0 if no cartridge is present) and you can now remove the short. The player will now stay in service mode until you turn the power off. If you do turn the player off at any time during the alignments, you will have to follow that procedure again if you want to put it back in service mode.

Different players use different procedures for placing them in service mode. Some players require that you leave a jumper in place; others require that you press and hold one or two buttons on the front panel when the unit is turned on. Like models use the same procedures.

Once in service mode you can manually switch different circuits on and off. Numbers will either be displayed, or in some cases not displayed on the front panel, indicating what test mode you are in.

The Laser Pickup

It is not uncommon to find that some players allow you to switch the laser off and on to see if it's functioning properly, or to adjust the output. If you do not have a laser power meter, checking the laser may be as simple as dimming the lights in the shop and cautiously looking at the lens from a distance of at least a foot, and preferably several feet, and at an angle to the direction of the beam. The beam can be easily seen at an angle quite a distance from the laser assembly.

Although the laser is fairly weak, observe proper precautions that are in the manual, and do not peer into it like a microscope. If you can see no light emitting from the laser (in a darkened room), and you are sure that the player is in the right test mode, there may be a problem with the laser.

Making the Adjustments

The first alignment on the Denon DCM 777 is the PLL adjustment. Before you can get the proper frequency, the test points ASY and GND3 will have to be connected with a jumper. The frequency you are looking for is 4.32MHz. If those two points are not connected together via a jumper, you will never even get close to the proper specs. If you are having problems getting 4.32MHz, and the adjustment is at its stop, or close to it, double check your jumper and connections. Also, this measurement is taken through a 10:1 scope probe, not only on the Denon, but on many other brands as well.

The next three adjustments: Tracking DC Offset, RF Offset, and Focus Offset are easily adjusted with a DVM. You can be as much as 50mV off on these adjustments, so as in horseshoes, close counts.

Tracking Offset is adjusted with a scope, using the filter that is called for. Again, this filter is easily constructed and is invaluable in seeing clear waveforms. Although you may have to deviate your scope settings slightly from what is called for, you should see a clear waveform just like the one pictured. Don't forget to switch your scope to dc.

A trick, given to me by Yamaha Technical Support, is to compress the wave into a simple vertical line (*Figure 3-3*). Although the normal waveform is clear, it is in constant motion and difficult to get a handle on. One vertical stripe is a snap to adjust.

Figure 3-3. The tracking offset waveform compressed into a single vertical line. Using this method makes it a snap to determine if it's symmetrical about the zero point.

Adjusting the Gains

Focus Gain is adjusted using a pair of filters, a frequency generator, a frequency counter, and a scope. A handful of small hook type connectors can be invaluable here. If you get confused as to where to inject the signal from the generator, once again remember the meaning of the O and the I; FEO (Focus Error OUT) and FEI (Focus Error IN). The generator goes to the input. It is usually best if you make the last connection the positive lead from the signal generator, with the player in the proper mode and the disc spinning.

Tracking Gain is adjusted almost exactly like Focus Gain, using the Lissajous waveform once again. The Denon manual gives scope settings for both Focus Gain and Tracking Gain. You will find that if your scope has V/div settings for both 10:1 and 1:1 probes, you should use the 1:1 setting with a 10:1 probe.

Virtually all alignment instructions will have you confirm Tracking Offset at some point. Once again, this is an easy alignment, especially if you remember to compress the waveform.

Denon completes their adjustments at this point, having you continue only if you are experiencing problems. Tangential adjustment, or aligning for the optimum Eye Pattern, is fairly simple. The Grating adjustment is not.

The Grating Adjustment

The Grating adjustment is probably the most difficult alignment to get right—on any player (*Figure 3-4*). And if this adjustment is not correct, the CD player will not play at all. Most new players can be adjusted with a simple screwdriver, but many cannot. Investing in costly alignment tools will greatly reduce your profitability, unless you can be assured they will be used again.

Second, the Grating adjustment is extremely difficult even with the right tools. The main problems are that the adjustment is minute, and that it is difficult to differentiate between the null point of the waveform, and the (exactly as it is translated) "Waveform of not null point."

The Grating adjustment aligns the beams in a "3 beam" laser assembly. A player that won't accept discs may only need a Grating adjustment, assuming the motor and laser circuits are operating properly. Low amplitude of this waveform may indicate a laser problem.

*Figure 3-4. The pointer points to the grating
adjustment on the Denon and similar multidisc players.
This type of player has the laser assembly aimed
downward. The disc is loaded information side up.*

Checking Your Work

When all alignments have been completed, turn off the power and remove
any jumpers and probes still attached. When you power up, the player should
return to normal operating mode. If you have one, a "torture disc" can be
used to check out your work.

A torture disc is one of those discs with built-in dirt spots, scratches, and
fingerprints. Philips manufactures a set of two discs, one with flaws, and one
without (for reference). Although pricey, around $100, these discs can be an
invaluable diagnostic tool.

If the customer complains of only one or two discs being a problem, make
sure they bring the discs in with the repair. If the discs are clean with no
major scratches, and you can duplicate the problem with these CDs, playing
them after the player has been serviced will help confirm that you have
indeed solved the problem, assuming the discs play fine on another player.

A Few Thoughts

You'd be amazed how many customers don't realize that the "business" side
of a CD is the side without writing on it. Yes, all their discs are clean, just on
the wrong side. Also, some players, like the Denon DCM 777, accept discs

upside down (music side up). Both the Denon and Pioneer cartridge type players share the same transport assembly, and load the same way. Although they are both very good products, neither will play a disc that has been loaded label side up.

Mechanical Problems

Remember the good old days of records (LPs)? Remember how you sometimes had to enlarge the hole to get the record to fit on the spindle? Well, some CDs have the same problem. They don't sit properly on the CD "turntable," and have to be gently reamed out. Since everything that goes on inside the player is hidden from view, you can't see that the disc is not sitting flat in the clamper assembly.

Waveforms that pulsate vertically may indicate a bent spindle. Waveforms that pulsate horizontally, like a "Slinky," may be an indication of a motor (speed) problem. And waveforms that will not reach proper amplitude may indicate some sort of laser problem.

You may want to check for service bulletins. The Denon DCM 777 had a problem with intermittent disc access. The problem is solved by replacing R144 (68K ohms) with a 91K resistor. This modifies the feedback circuit driving the spindle motor, allowing the CD to "spin up" faster. If the problem still persists, a new motor may be required.

This Information is Portable

Alignments on other models are very similar, although the labels and order of adjustments may not be the same. For example, on a particular Yamaha player the PLL adjustment is called VCO Freerun. Yet, you are still looking for a reading of 4.32MHz, and you still are required to use a 10:1 probe.

The more you do these adjustments the easier they get. You will soon begin to see the similarities from one brand to the next, and when you come across one of those really vague service manuals, your previous experience will go a long way.

Experience will also teach you what waveforms to look for when evaluating problems, regardless of the manufacturer or the absence of a manual. It's a great feeling to be able to tell your customers what the problem is, without having to repair the unit first.

Although proper alignment will add time to any repair, I'm reminded of an old saying, "We can never seem to find the time to do it right, but we always have the time to do it over again."

Chapter 4

CD Player Fundamentals—Part I

By Sam Goldwasser

Compact disk players are complex because of the digital decoding of the music that is encoded on the disk. This process is complex, but many of the faults that occur in CD players are relatively simple to diagnose and repair. Part I of this two part article describes the construction of CD players.

Power Supply

CD players that are designed to be used in component stereo systems normally feature linear power supplies. These supplies are reliable and easy to fix. Portable CD players are likely to use switching supplies, possibly sealed in a shielded can. These can be difficult to repair.

Usually, at least three voltages are needed for the circuits in a CD player: logic power (e.g. Vcc of +5V) and a pair of voltages for the analog circuitry (e.g., +/- 15V). However, some designs use a variety of voltages for various portions of the analog (mainly) circuitry.

Electronics Board

The electronics board in a CD player contains the microcomputer controller, servos, readback electronics, audio D/A(s) and filters. Most servo adjustment pots will be located on this board. In many cases these adjustments are clearly marked, but not always. Do not turn any adjustment controls unless you are sure of what you are doing, and then only after marking their original positions precisely.

The Optical Deck

The optical deck includes all of the components required to load and spin the disc, the optical pickup, and its positioning mechanism:

• Loading drawer—Most portable and many lower cost CD players lack the convenience of a loading drawer. Most loading drawers are motor driven. However, some must be pushed in or pulled out by hand.

Common problems: loose or oily belt causing drawer to not open or close, or to not complete its close cycle. There can be mechanical damage such as worn/fractured gears or broken parts. The drawer switch may be dirty causing the drawer to decide on its own to close. The motor may be shorted, have shorted or open windings, or have a dry or worn bearing.

• Spindle/spindle table—When the disk is loaded, it rests on this platform which is machined to automatically center it and minimize runout and wobble.

Common Problems: Dirt on table surface, bent spindle, dry or worn bearings if spindle not part of motor but is belt driven, loose spindle.

• Spindle motor—The motor that spins the disk. Most often the spindle platform is a press fit onto the spindle motor. Two types are common: The first is a miniature dc motor (using brushes) very similar to the common motors in toys and other battery operated devices. The second type is a brushless dc motor using Hall effect devices for commutation. In very rare cases, a belt is used to couple the motor to the spindle.

Common problems: partially shorted motor, shorted or open winding, dry/worn motor bearings.

• Clamper—The clamper is usually a magnet on the opposite side of the disk from the spindle motor which prevents slippage between the disk and the spindle platform. The clamper is lifted off of the disk when the lid or drawer is opened. Alternatively, the spindle may be lowered to free the disk.

Common problems: Clamper doesn't engage fully, permitting disk to slip on spindle due to mechanical problem in drawer closing mechanism.

• Sled—The sled is the mechanism on which the optical pickup is mounted. The sled provides the means by which the optical pickup can be moved across the disk during normal play or to locate a specific track or piece of data. The sled is supported on guide rails and is moved by either a worm or ball gear, a rack and pinion gear, linear motor, or rotary positioner similar to that used in a modern hard disk drive. This list is in increasing order of performance.

Common problems: dirt, gummed up or lack of lubrication, damaged gears.

• Pickup motor—The entire pickup moves on the sled during normal play or for rapid access to musical selections. The motor is either a conventional miniature permanent magnet DC motor with belt or gear with worm, ball, or rack and pinion mechanism, or a direct drive linear motor or rotary positioner with no gears or belts.

Common problems: partially shorted motor, shorted or open winding, dry or worn bearings.

• Optical pickup—This unit is the 'stylus' that reads the optical information encoded on the disk. It includes the laser diode, associated optics, focus and tracking actuators, and photodiode array. The optical pickup is mounted on the sled and connects to the servo and readback electronics using flexible cables.

Common problems: hairline cracks in conductors of flexible cable causing intermittent behavior.

Components of the Optical Pickup

All of the parts described below are in the optical pickup. As noted, the optical pickup is usually a self-contained and replaceable subassembly. While optical pickups are precision optomechanical devices, they are re-markably robust in terms of susceptibility to mechanical damage.

• Laser diode—The laser diode emits infrared (IR) light, usually at 780nm. This is called the "near IR," just outside the visible range of 400nm to 700nm. The power output of the diode is no more than a few mW. The power level of this beam is reduced to 0.25mW to 1.2mW at the output of the objective lens. A photodiode inside the laser diode case monitors optical power directly and is used in a feedback loop to maintain laser output at a constant and extremely stable value.

Common problems: bad laser diode or sensing photodiode resulting in reduction or loss of laser output.

• Collimating lens—The collimating lens converts the wedge shaped beam of the laser diode into nearly parallel rays.

• Diffraction grating—In a "three-Beam pickup," the diffraction grating generates two additional lower power (first order) beams, one on each side of the main beam, which are used for tracking feedback. There is no diffraction grating in a "single-beam pickup."

• Cylindrical lens—In conjunction with the collimating lens, the lens provides the mechanism for accurate dynamic focusing by changing the shape of the return beam based on focal distance.

• Beam splitter—This device passes the laser output to the objective lens and disk and directs the return beam to the photodiode array.

• Turning mirror—Redirects the optical beams from the horizontal of the optical system to the vertical to strike the disk.

Common problems: dirty mirror. Unfortunately, this may be difficult to access for cleaning.

The previous four items are the major components of the fixed optics. Outside of damage caused by a serious fall, there is little that can go bad in this subassembly. Better hope so in any case—it is usually very difficult to access the fixed optics components and there is no easy way to realign them anyhow. Fortunately, except for the turning mirror, it is unlikely that they would ever need cleaning. Usually, even the turning mirror is fairly well protected and remains clean.

• Objective lens—The objective lens is a high-quality focusing lens, very similar to a good microscope objective. This lens has an N.A. of 0.45, and a focal length of 4mm. It is made of plastic with antireflection coating (the blue tinge in the center).

Common problems: dirty lens, dirt in lens mechanism, damage from improper cleaning or excessive mechanical shock.

• Photodiode array—This is the sensor which is used to read back data and control beams.

Common problems: bad photodiode(s) resulting in improper or absence of focus and weak or missing RF signal.

• Focus actuator—Since focus must be accurate to 1 micron (1μm), a focus servo is used. The actuator is actually a coil of wire in a permanent magnetic field like the voice coil in a loudspeaker. The focus actuator can move the objective lens up and down; closer to or farther from the disk based on focus information taken from the photodiode array.

Common problems: broken coil, damaged suspension (caused by mechanical shock or improper cleaning techniques).

• Tracking actuator—Like focus, tracking must be accurate to 1μm or better. A similar voice coil actuator moves the objective lens from side to side based on tracking feedback information taken from the photodiode array.

Note: on pickups with rotary positioners, there may be no separate tracking coil, as its function is subsumed by the positioner servo. The frequency response of the overall tracking servo system is high enough that the separate fine tracking actuator is not needed. Common problems: broken coil, damaged suspension (caused by mechanical shock or improper cleaning techniques).

Classification of CD Player Problems

While there are a semi-infinite number of distinct things that can go wrong with a CD player, symptoms can be classified as a hard failure or a soft failure.

A hard failure is one that causes the unit to fail to operate at all, such as door opening/closing problems, disk is not recognized, no sound, or the unit totally dead.

A soft failure is one in which the unit operates in some fashion, but improperly, such as skipping, continuous or repetitive audio noise, search or track seek problems, or random behavior.

Both of these types of problems are common. The causes in both cases are often very simple, easy to locate, and quick and inexpensive to repair.

Most Common CD Player Problems

While it is tempting to blame the most expensive component in a CD player, the laser, for every problem, this is usually uncalled for.

Here is a short list of common causes for a variety of tracking and audio or data readout symptoms:

• Dirty optics—This includes the lens, prism, or turning mirror.

• Drawer loading belts—worn, oily, flabby, or tired.

• Sticky mechanism—dirt, dried up/ lack of lubrication, dog hair, sand, etc.

• Broken (plastic) parts—gear teeth, brackets, or mountings.

• Need for electronic servo adjustments—focus, tracking, or PLL.

• Intermittent limit or interlock switches—worn or dirty.

• Bad connections— solder joints, connectors, or cracked flex cable traces.

• Motors—electrical (shorted, dead spot) or mechanical (dry/worn bearings).

• Laser—dead or weak laser diode or power problems.

• Photodiode array—bad, weak, or shorted segments or no power.

• Bad/heat sensitive components.

Most Frequent Problem Areas

The following two areas cover the most common types of problems you are likely to encounter. For any situation where operation is intermittent or audio output is noisy, skips, or gets stuck, or if some disks play and others have noise or are not even recognized consistently, consider these first:

• Dirty lens—This problem is especially likely if the location in which the player is used is particularly dusty, the player is located in a greasy location like a kitchen, or there are heavy smokers around. Cleaning the lens is relatively easy and may have a dramatic effect on player performance.

• Mechanical problems—dirt, dried up lubrication, damaged parts. These may cause erratic problems or total failure. The first part of a CD may play but then get stuck at about the same location.

If your CD player has a "transport lock" screw, check to see that it is turned to the "operate" position.

General Inspection, Cleaning, and Lubrication

The following should be performed as general preventive maintenance or when any erratic behavior is detected. The lens, drawer mechanism, and sled drive should be checked, and cleaned and/or lubricated if necessary.

You will have to get under the clamp to access the lens (drawer loading models). Be gentle. No lubrication is needed, and none should be used anywhere in the lens assembly.

At the same time, you will have access to the spindle.

• Objective lens—Carefully clean the lens assembly. Be careful! The lens is suspended by a voice coil actuated positioner which is relatively delicate. A CD lens cleaning disk is nearly worthless except for the most minor dust, as it will not completely remove grease, grime, and condensed tobacco smoke products (yet another reason not to smoke), and the disk may make matters worse by just moving the crud around.

First, gently blow out any dust or dirt which may have collected inside the lens assembly. A photographic type of air bulb is fine but be extremely careful using any kind of compressed air source. Next, clean the lens itself. It is made of plastic, so don't use strong solvents. There are special cleaners, but alcohol (91% medicinal is acceptable, pure isopropyl is better. Avoid rubbing alcohol especially if it contains any additives) works fine for CD players and VCRs.

There should be no problems as long as you dry everything off (gently) reasonably quickly. Do not lubricate! You wouldn't oil a loudspeaker, would you?

When the lens is clean, it should be perfectly shiny with a blue tinge uniform over the central surface. If you can get to the turning mirror or prism under the lens without disturbing anything, clean that as well using the same procedure.

Do not use strong solvents or anything with abrasives—you will destroy the lens surface most likely rendering the entire expensive pickup worthless.

It is easy to be misled into thinking that there are much more serious problems at the root cause of disks not being recognized, audible noise, and tracking problems like skipping, sticking, or seek failures. However, in many cases, it is simply a dirty lens.

Spindle Bearing

Check the spindle bearing (this is primarily likely to cause problems with repetitive noise). There should be no detectable side to side play, i.e., you should not be able to jiggle the platform that the CD sits on. If you find that the bearings are worn, it is possible to replace the motor (about $10 from various mail order houses), though removing and replacing the disk platform may prove challenging as a result of the usual press fit mounting.

The focus servo can compensate for a vertical movement of the disk surface of 1mm or so. A small bearing side play can easily cause larger vertical errors—especially near the end (outer edge) of the disk. Even if you are not experiencing problems due to bearing wear, keep your findings in mind for the future.

On some players there is a bearing runout adjustment screw on the bottom of the spindle if the spindle is not driven by a standard permanent magnet motor. I have seen this in a Sony Discman which had a custom motor assembly. A small tweak to this adjustment may correct a marginal spindle problem.

To access the drawer mechanism and sled drive in component units, you will probably need to remove the optical deck from the chassis. It is usually mounted by 3 long screws (one of which may have a grounding tab. Don't lose it. In portables, the bottom panel of the unit will need to be removed. Try not to let any of the microscrews escape! A good set of jeweler's screwdrivers is a must for portables.

Drawer Mechanism

(If present)—Check for free movement of the drawer mechanism. Test the belt for life—it should be firm, reasonably tight, and should return to its original length instantly if stretched by 25% or so. If the belt fails any of these criteria, it will need to be replaced eventually, though a thorough cleaning of the belt and pulleys with isopropyl alcohol (dry quickly to avoid damaging the rubber) or soap and water may give it a temporary reprieve.

Also, check the gears and motor for lubrication and damage and correct as necessary. Clean and lubricate (if necessary) with high quality light grease suitable for electronic mechanisms such as moly lube or silicone grease. A drop of light oil (electric motor oil, sewing machine oil) in the motor bearings may cure a noisy or dry bearing.

Sled Drive

Check the components that move the pickup including (depending on what kind of sled drive your unit has) belt, worm gear, other gears, slide bearings. These should all move freely (exception: if there is a lock to prevent accidental damage while the unit is being transported the pickup may not move freely or very far). Inspect for damage to any of these components which might impede free movement. Repair or replace as appropriate. Clean and lubricate (if necessary) with just a dab of high quality light grease suitable for electronic mechanisms such as moly lube or silicone grease). A drop of light oil (electric motor oil, sewing machine oil) in the motor bearings may cure a noisy or dry bearing.

Try to play a disk again before proceeding further. I guess you most likely have already done this.

Lubrication of CD Players

The short recommendation is: do not add any oil or grease unless you are positively sure it is needed. Most moving parts are lubricated at the factory and do not need any further lubrication over their lifetime. Too much lubrication is worse then too little. It is easy to add a drop of oil but difficult and time consuming to restore an optical pickup that has taken a bath in lubricant.

Never use any type of lubricant that is not expressly recommended by the manufacturer of the CD player. This includes any of the highly touted name brand lubricants on the hardware store shelves.

A light machine oil like electric motor or sewing machine oil should be used for gear or wheel shafts. A plastic safe grease like silicone grease or moly lube is suitable for gears, cams, or mechanical (piano key) type mode selectors. Never use oil or grease on electrical contacts.

Unless the unit was not properly lubricated at the factory (which is quite possible), don't add any unless your inspection reveals the specific need. In a CD player, there are a very limited number of failures that are caused by lubrication.

Note that in most cases, oil is for plain bearings (not ball or roller) and pivots while grease is used on sliding parts and gear teeth. If the old lubricant is gummed up, remove it and clean the affected parts thoroughly before adding new lubricant, oil or grease.

In general, do not lubricate anything unless you know there is a need. Never 'shotgun' a problem by lubricating everything in sight! You might as well literally use a shotgun on the equipment!

Part II of this article will discuss some typical malfunctions associated with the CD player and will suggest some corrective measures you can take to fix them.

Chapter 5

CD Player Fundamentals—Part II

By Sam Goldwasser

Part 1 of this article, which appeared in the October issue, provided a general overview of CD player operation, including some general comments on the types of problems that can occur. This segment includes a discussion of some specific problems that may cause a CD player to malfunction, or to fail to operate entirely, and provides specific suggestions for corrective action.

Where To Start if the Player is Totally Dead

If a CD player is totally dead, check input power, power cord, fuse, and power supply components. Locate the outputs of the power transformer and trace them to the rectifiers and associated filter capacitors and regulators. While the actual voltages will probably not be marked, most of the power in a CD player will be typically between +15Vdc and -15Vdc. Sometimes the voltage ratings of the filter capacitors and/or regulators will provide clues as to correct power supply outputs. Don't forget the obvious: the line cord, line fuse (if present), and power switch, or outlet. Most component CD players use linear power supplies so troubleshooting is straightforward. However, portables use dc-to-dc converters to generate the several voltages required. These are more difficult to troubleshoot. If an incorrect power adapter was used, then major damage can result despite the various types of protective measures taken in the design.

I inherited a Sony Discman from a guy who thought he would save a few bucks and make an adapter cord to use it in his car. Not only was the 12V to 15V from the car battery too high, but he got it backwards. This blew the dc-to-dc converter transistor in two despite the built-in reverse voltage protection and fried the microcontroller. Needless to say, the player was a loss but the cigarette lighter fuse was happy as a clam.

The moral of this story is that those voltage, current, and polarity ratings marked on portable equipment are there for a reason. Voltage ratings should not be exceeded, though using a slightly lower voltage adapter will probably cause no harm, performance may suffer. The current rating of the adapter should be at least equal to the printed rating. The polarity, of course, must be correct.

If the power is connected backwards with a current limited adapter, there may be no immediate damage depending on the design of the protective circuits. But don't take chances; double check that the polarities match before you plug it in.

Note that even some identically marked adapters put out widely different open circuit voltages. If the unloaded voltage reading is more than 25% to 30% higher than the marked value, I would be cautious about using the adapter without confirmation that it is acceptable for your player. Needless to say, if the player behaves in any strange or unexpected manner with a new adapter, if any part gets unusually warm, or if there is any unusual odor, unplug it immediately and attempt to identify the cause of the problem.

CD Player is Operational But There is No Display

If the CD player operates but the display is blank, the display may be one that requires backlighting, which uses miniature incandescent lamps. The lamp may be burned out. If you have to replace a burned-out lamp from a CD player display, check to see if you can find an alternative to the high-priced exact replacement bulbs. Test the bulbs with an ohmmeter. Measure the voltage across the light bulb connections and then replace the bulb with one that is specified at about 25% to 50% higher voltage. These may not be quite as bright but should last forever.

If the light bulbs are not at fault, or if there are no light bulbs, then check for power to the display including bad connections or connectors that need to be reseated. There could also be a power supply (e.g., missing voltage for a vacuum fluorescent display) or driver problem.

CD Player Ignores You

Symptoms like the display coming up normal when the power is turned on but all (or certain) commands are ignored could mean any of several things:

• Front panel problem—one or more buttons are not responding. Reseat internal cables, clean or replace offending push button switches. If your CD player has a remote control, see if it operates the player correctly.

• Reset failure—the player has failed to reset properly and is not ready for user input. Check power supply voltages, reseat internal connectors.

• Controller and/or driver electronics for the affected functions are defective. Check power supply voltages, reseat internal connectors.

For all but the first one, a service manual will probably be needed to proceed further if the problem is not caused by a bad power supply or bad connections.

Drawer Does Not Open Or Close

If the drawer doesn't open when the front panel button is pressed, listen for the motor attempting to open the drawer. If you hear it whirring but nothing happens, check for an oily/loose belt or other mechanical failures. Cleaning of the belt may provide a temporary repair, but it should be replaced for a proper repair.

If you don't hear any activity from the loading drawer motor, the problem could be caused by the motor, the control chip, or the front panel pushbutton. Try operating the player by using the remote control to determine if the problem is caused by the pushbutton.

Drawer Operation is Erratic

You are about to remove your favorite CD but the player beats you to it, closes the drawer, and starts playing it over again. Or the drawer reverses course halfway out. Or it may be that the drawer motor continues to whir even after the door is fully open or closed and the front panel is then unresponsive.

This is usually due to dirty contacts on the door position sense switches. There are usually 3 sets of switch contacts associated with the drawer mechanism. If any of these get dirty, worn, or bent out of place, erratic operation can result:

• Drawer closed sense switch—dirty contacts may result in the drawer motor continuing to whir after the door closes and the front panel may then be unresponsive. Eventually, the drawer may open on its own.

• Drawer open sense switch—dirty contacts may result in the drawer motor continuing to whir after the door opens and the front panel may then be unresponsive. Eventually, the drawer may close on its own.

• Drawer pushed sense switch—most CD players allow the user to start play by gently pushing on the drawer which depresses a set of switch contacts. If these are dirty, the result may be that the drawer decides to close on its own or reverses direction in the middle of opening or closing operation.

The solution to all these problems is usually to simply locate the offending switches and clean their contacts. These switch contacts are usually not protected from dust, dirt, and grime so that these types of problems are quite common.

Drawer Does Not Close Completely

Failure of the drawer to close is a symptom that may not be obvious. The drawer may appear to close but a loose or oily belt may prevent the mechanism from completing the close cycle. This can result in erratic behavior because the disc clamping action is often controlled by the movement of the drawer.

The result of this problem is that sometimes the player will not recognize the disc, sometimes the drawer will open, or the problem may cause more subtle failures like tracking problems, etc. If you suspect that a CD player is experiencing this problem, clean the belt and see if there is any improvement. Belt replacement will be necessary eventually. Check for gummed up lubrication as well.

If the drawer goes through the motions of closing and then stops short without any further sounds, a gear may have jumped a tooth or broken some.

The result is either that the mechanism is now incorrectly timed or not able to complete the operation. Examine the mechanism closely for broken parts. Cycle it manually by turning the appropriate motor pulley or gear to see if the drawer gets hung up or is much more difficult to move at some point.

If the player continues to make a whirring sound after the drawer stops, there might be some other kind of mechanical damage resulting in an obstruction or really gummed up lubrication not allowing the operation to complete.

Spindle Table Loose or Sticks To Clamper Upon Eject

When you remove the CD, you may have an added surprise—the platform upon which the CD sits pops off as well, possibly jamming everything. There may also be startup and spindown problems.

Various models use different techniques to fasten the spindle table to the motor shaft but this is strictly a mechanical problem. Either a set screw has worked loose, adhesive has weakened, or a press fit has come undone.

If there is no set screw, a drop of epoxy may be what is needed. However, correct height is important to guarantee proper focus range so some care will be needed if there is no definite stop. The disc and rotating clamper magnet must be clear of any fixed structures and at the correct distance from the optical pickup. Where something irreversible is involved like adhesive, check the service manual. The specification is usually 0.1mm accuracy.

A loose spindle table may also result in continued spinning upon eject or sluggish or noisy startup or seek, because if the spindle is loose, the motor will not be able to properly control disc speed during speed changes.

Intermittent Operation

When a CD player appears to have a mood problem; playing fine sometimes or for only part of a disc or aborting at random times, there can be several possible causes including a dirty lens, dirty or worn interlock switch or bad connections to the interlock switch (mainly in portables and boomboxes), flexible cable with hairline cracks in one or more conductors, other bad connections, marginal power supply, or a defective disc.

• Dirty, scratched, or defective CD—confirm that the CD is not the problem. Clean the disc and/or try another one.

• Dirty lens—a player that accepts some discs and not others or accepts discs sporadically may simply need its eyeglasses cleaned.

• Mechanical—oily, flabby belts preventing full drawer closing or gummed up lubrication on the sled (may fail depending on ambient temperature. For example, if the music gets stuck at about the same time on every disc, then there may be gunk on the end of the sled track preventing the sled from moving any farther. This is especially likely if you just purchased a disc with an unusually long playing time—it has nothing to do with the musical tastes of the CD player!

Note: some players will simply not play discs which exceed about 74 minutes—the legal limit for CD playing time. Some discs may be as long as 78 minutes or more which means that some aspects of the CD specifications were compromised.

• Bad connections—there are often many little connectors used to get signals and power between the optical deck and main circuit board. These are usually cheaply made and prone to failure. Wiggling and reseating these may cure these problems. There may even be bad solder connections on the pins of connectors or board mounted switches. Slight flexing or just expansion and contraction may result in intermittent shutdown or other problems. These problems are more likely with portables and boomboxes which are subject to abuse.

• Cracks in ribbon cable—The moving and fixed parts of the optical pickup are often joined with a printed flexible cable. Constant flexing may cause one or more of the copper traces to crack. This may show up as an inability to get past a certain point on every disc—the player may shut down or start skipping at around 23 minutes into every CD.

• Dirty switches—oily film or oxidation may be preventing any of the limit or interlock switches from making reliable contact. If this is the case, the player may stop at random times, fail to accept a disc, close the drawer without your permission, etc. Use contact cleaner and typing paper to clean the contacts. Disassembly may be required for enclosed switches.

• Power supply or logic problems are also possible but rare. However, if you have a scope, check the power supply outputs for ripple; a filter capacitor may have dried up and lost most of its capacitance.

CD Player Overheats

A CD player which becomes noisy may have a component that is heating up and changing value.

In general, there should not be much change in behavior from the instant the power is applied until the next millennium. There is not much in a CD player that runs hot and might change characteristics. However, components do sometimes fail in this manner. Problems of this type need to be diagnosed in much the same way as one would find overheating components in a TV or computer monitor.

You will need a can of cold spray ('circuit chiller') and an oscilloscope, if available. Even a hair dryer on the no-heat setting will work in a pinch.

You are going to have to try cooling various components to try to determine which one is bad. However, on a unit that dies completely and suddenly after it warms this will not be much fun since you will not have ample opportunity to detect changes in behavior. On a CD player that will play but with tracking problems and/or audio noise, you should be able to monitor the playback quality by simply listening for improvement when you have cooled the flaky part.

First, I would recommend running with the covers removed and see if that has an effect confirming a thermal problem. Next, use the cold spray on individual components like the LSI chips—quick burst, wait a few seconds for something to change. If you are using the hairdryer, make a funnel out of paper to direct the air flow. You will need to be more patient with this approach.

Another way to determine if the problem is caused by heat is to use the scope to look at the RF "eye" pattern during this time and see if it decreases in amplitude and/or quality over the course of an hour. If it does, you may have an overheating problem in the laser diode or its power supply.

Chapter 6

More Musical Instrument Amps and Other MI Equipment

By Ron C. Johnson

Those of you who have been around the service business for a while are probably not surprised that tubes are still used extensively in guitar amps. And you've probably seen your share of tube circuits from the "old days" of television. On the other hand, some of you younger guys may find the idea pretty archaic. But, really, we're still using vacuum tubes as CRT's in televisions and monitors, so we're not all that far removed from the days of vacuum tubes.

Servicing Vacuum-tube Amplifiers

As I mentioned in a previous article, many guitar amps still use tube circuits because musicians like the sound they get out of them. Solid state amps just seem to be too clean. Not enough texture, some would say. Whatever the reason, if you find yourself repairing musical instrument electronics you're going to run into tube amps.

Normally, the first thing I would say about tube amps is to be careful. Four hundred volt power supplies require some care and attention to avoid a shock. But anyone who works on televisions on a regular basis should already be aware of the danger of high voltage. Just the same, don't let your guard down just because you're working on an amplifier instead of a TV. The results of an electrical shock can run from mild discomfort all the way to electrocution and the dangers of secondary injuries can be just as serious.

A Typical Tube Amp

The circuit shown in *Figure 6-1* is a typical low power bass guitar amp made by Fender. This one only uses four tubes: two 7025's (dual triodes) and two 6V6GTA's (pentodes). Both of these are very common (or at least used to be when tubes were in general use). The 7025 is used as a signal amplifier and the 6V6's are power outputs. For those of you unfamiliar with tube circuits, a few quick points follow.

Vacuum tubes work on the basis that, if you apply a potential between two metal electrodes in a vacuum, and then provide a source of free electrons (by the heating one of the electrodes), electrons will flow through the vacuum from the negative (heated) electrode to the positive.

The positive electrode is called the anode (or plate) and the negative electrode is called the cathode. The heater, or filament, is under or inside the cathode. It heats the cathode, releasing electrons which are attracted to the plate.

The Vacuum-tube Diode

A tube with two electrodes constitutes a diode. If the plate is positive and cathode negative, and there is a closed circuit between the two, external to the tube, current will flow. Within the tube, this current will be from positive to negative if you use conventional current flow, or vice versa if you use electron flow.

I personally use conventional current flow although some still prefer the other. If the voltage is applied in the opposite polarity no current flows. Because the current flows if the voltage is applied in one direction, but doesn't flow when the voltage is in the opposite direction, the vacuum-tube diode is a rectifier.

For amplification, vacuum tubes add at least one more electrode, (called a grid), between the anode and cathode. By applying a voltage to the grid, the current flow from anode to cathode can be controlled. If you're familiar with FET's you can easily catch on to tubes. A voltage on the gate (grid) controls the current flow (transconductance) of the device from drain (anode) to source (cathode).

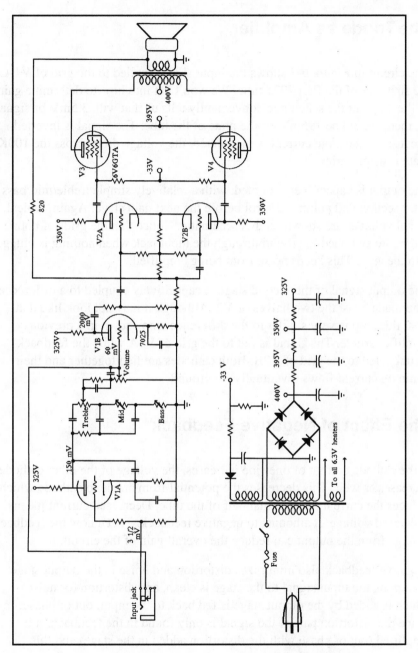

Figure 6-1. *This is a schematic diagram of the circuit of a typical low power bass guitar amp made by Fender. This one only uses four tubes: two 7025's (dual triodes) and two 6V6GTA's(pentodes).*

The Triode as Amplifier

The circuit in *Figure 6-1* shows the input signal applied to the grid of V1A, the first half of the first 7025 tube. We won't get into how to determine gain of the amp but the schematic conveniently tells us that with 3.5mV of signal in, there should be 195mV on the plate of the tube. The signal is inverted because, as the plate current is modulated, the voltage drop across the 100K plate resistor varies.

The output is capacitively coupled, with a relatively simple treble/mid/bass filter section and volume control before the next gain stage. Again, typical signal voltages are shown on this schematic. Notice that the grids of both stages are switched to ground through the input jack when nothing is plugged into the amp. This keeps noise from being amplified.

The output signal of the second stage is capacitively coupled to a difference amp made up of the two halves of V2. Although it may not look like it at first, this amp is very similar to the difference amp shown in a previous amplifier article. The signal is fed to the grid of V2A while the feedback signal is fed to the grid of V2B. Both cathodes are tied together and their common current flows (eventually) to ground.

The Effect of Negative Feedback

If the cathode current of one side increases, the voltage of the other cathode increases as well. This decreases the potential from anode to cathode which reduces the current through that half of the tube. Decreased current means decreased voltage. It amounts to negative feedback. That's how the feedback voltage from the output can reduce the overall gain of the circuit.

Negative feedback also minimizes distortion and noise in the output stage. Assuming the input signal to the stage is clean, any distortion (or noise) which is added by the output stage is fed back to the input, out of phase. Since the distortion part of the signal is only found in the feedback, it is amplified (out of phase with the distortion added in the stage) and this subtracts out, leaving a clean output.

Push-pull Output

The output stage is a push-pull configuration: one side drives the output on the positive excursion of the signal while the other side takes care of the negative excursion. Notice the additional -33V power supply which sets a bias on the grids of each output tube.

The 6V6 pentode is a power tube. Its extra internal grids give it better drive capacity without the capacitive and grid limitations of the smaller tubes. In this case the 400V power supply is connected to the center tap of the output transformer. On each signal excursion current flows in alternate directions through the input windings inducing a current in the output windings which are connected to the speaker.

The transformer not only allows this bias arrangement but it matches the high impedance tube output circuit to the low impedance speaker load. The feedback signal is obtained from the output side of the transformer and sent back to the input of the difference amp to control the gain.

Servicing

Tube circuits are actually pretty easy to troubleshoot. There are two dc power supplies here: the 400V main supply and the -33V bias supply. The 400V supply uses a full wave bridge while the negative supply uses a single diode for half-wave rectification. Each has its own transformer winding.

Watch for bad power supply caps which will cause large levels of ripple on the supply rails. Also, make sure the tube filament voltage is present. This 6.3Vac is provided by a separate output winding on the power supply transformer and is connected to each tube.

Tubes, of course, can fail. They tend to be somewhat fragile, especially in an amp that is hauled around to gigs. Sometimes the internals come loose and the tube goes "microphonic". Even though it amplifies, the slightest vibration will be amplified as well. In extreme cases the vibrations from the speaker itself will feed back causing the amp to howl.

The positive aspect of tubes is that they are easily replaced and can be tested if you have a tester available. I wouldn't trust a tube tester too far though.

Signal Tracing

Signal tracing through a circuit like this is fairly easy as well. You need to know which pin is which on the tubes. I have an old GE tube manual which lists most of the common tubes, their specs and pinouts. Schematics like this one give typical signal levels which is helpful. Watch out for open capacitors, shorted or open rectifiers in the power supply, and dirty, intermittent potentiometers in the volume and tone control circuits. Also look for signs of abuse. It's fairly common to find dried residue of beer and soft drinks inside the chassis.

The newer tube amplifiers have tube sockets mounted on printed circuit boards. The older ones still have point to point wiring between sockets. These can be a nightmare to troubleshoot and worse when you're soldering or desoldering components. Usually, the easiest way to remove a component is to cut it out and then remove the solder and the ends of the component leads.

Well, that's a pretty sketchy overview of a typical tube amp but hopefully it will help somebody to feel more comfortable tackling one for the first time.

Mixing Boards and Effects

Sound mixing boards are another kind of musical instrument equipment that you'll see quite often probably because they receive a fair amount of use and abuse. They run from small four channel microphone mixers to large road consoles with 32 or more input channels, monitor and effects mixes, and several submixes.

The features on these are limitless. Some have built in "phantom power" (a technique of simplexing dc power out to condenser microphones), clipping indicators, LED bargraphs, solo and cue switching, and so on. Ultimately, they all do basically the same thing. They take audio signals and route them to various outputs while controlling their levels.

One of the common problems with mixers is dirty or damaged potentiometers. Each channel has a volume "fader" or linear potentiometer which adjusts the main output level of that channel. Each channel will have several other rotary potentiometers which adjust equalization (tone) in several ranges, signal level to monitor and effects busses, input signal trim controls (to avoid overdriving the input circuitry), and other controls depending on the complexity of the board.

Cleaning Noisy Controls

Inevitably you'll have to clean up noisy potentiometers using spray cleaners. I recommend you find the best kind available and use it liberally. Unfortunately, these pots wear out and eventually cleaning them won't help. You'll have to replace them. Also I've only had limited success trying to clean linear faders. These have to operate smoothly and can't create any noise.

Usually trying to clean faders results in a quiet but sticky action which the customer won't accept. Keep a good stock of replacement faders for the brand of mixer you repair. They are expensive but necessary to do the job right. You'll also find that, often, these faders can only be obtained from the manufacturer of the equipment as they have been specifically manufactured for that product.

Locating Noisy Components

Another common problem with mixers is noise generated somewhere on the board. This can be caused by bad filter caps in the power supply, leaky signal caps or leaky transistors or op amps. The frustrating part of troubleshooting these problems is in trying to find the source of the problem.

A heat gun and some freeze spray can be handy here as some noise problems are caused by thermal defects in components. Warm up the suspected area (not too hot) and then spray individual components. Sometimes this will show up a bad one. Dirty connections and cracked circuit boards are common sources of problems here as well. Just moving the equipment around, or flexing the printed circuit board, will show up some problems.

Although the newer (and more expensive) mixers are being engineered for easier servicing, you'll find half the job on some repairs is just getting the thing open. Besides removing lots of screws you often have to remove all the knobs and nuts from the potentiometers before you can get to the solder side of the printed circuit board. For one particular brand and model I used to have a special, home-made "puller" just to get the knobs off without damaging the mixer.

Finally, don't always believe what the customer tells you about a mixer. Mixers are complex pieces of equipment and sometimes the problems are caused by the operators themselves. So insist that the customer gives a detailed explanation of the problem and then check it carefully yourself before tearing the board apart to start testing.

Chapter 7

Now Hear This

By John S. Hanson

Are you tired of ho-hum repairs and need something profitable to perk up business? Read on. Many of the new large-screen direct-view and projection TV sets have decent audio systems that you can customize to improve the sound of your customers' home entertainment systems and make some money in the process.

To determine if any given TV set has a controllable audio output, look in the back and see if it has a Jack Pack. The Jack Pack is a TV sales feature that makes the TV think it's a monitor. Simply put, it's a way of getting audio and video into and out of a TV set. The feature is seldom used and most owners of these sets are not aware of the feature unless a sales person calls it to their attention. This is where you enter the scene.

Improving the Audio Experience

Zenith, for one, offers several options on their Jack Packs that can actually expand the customer's sound and video experience. I will not dwell on the benefits of surround sound, but will confine my dissertation to what's available beyond the TV's internal speaker system. While the quality of sound that those speakers can produce is amazing, the sound can be even better.

Zenith's Jack Pack has two pairs of jacks that provide audio output (*Figure 7-1*). One pair, labeled "fixed audio," has line level audio suitable for input to an external stereo amplifier. In this mode, the remote volume function is inoperative, and the stereo amplifier's controls take over.

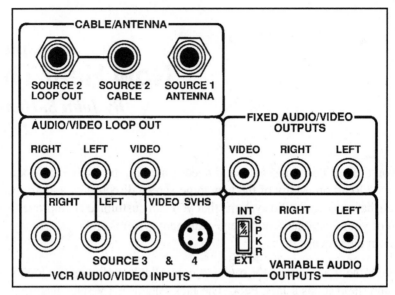

Figure 7-1. This jack pack has two pairs of jacks that provide audio output. The pair labeled "fixed audio," has line level audio suitable for input to an external stereo amplifier.

If you change the audio setup of any TV installation, be careful about the placement of the amplifier's speakers. It's possible that the TV sound may appear to be out of sync with the video. This can happen if the speakers are too far away from the TV screen. As a matter of reference, movie theater main channel speakers are always located directly behind the porous movie screen.

A Sound Idea

As with many other individuals who are "out of warranty", I am beginning to miss some words and say "what?" a lot. Many people with this problem are still in the denial stage and haven't considered hearing aids. Enhanced TV audio can be a boon to people with this affliction.

On my big screen Zenith, I noted that the external audio jacks on the Jack Pack were marked "Variable Audio." I assumed that this output would be under full control of the remote. Checking the schematic (*Figure 7-2*), I noted the internal/external speaker switch and the isolation and impedance matching network, understandable as the source is the full output of the audio amplifier.

Curious, I connected two four-inch 8ohm car stereo speakers to the jacks and flipped the speaker switch. With the remote I ran up the TV volume level to see if there was enough audio for what I had in mind. Now comes the good part.

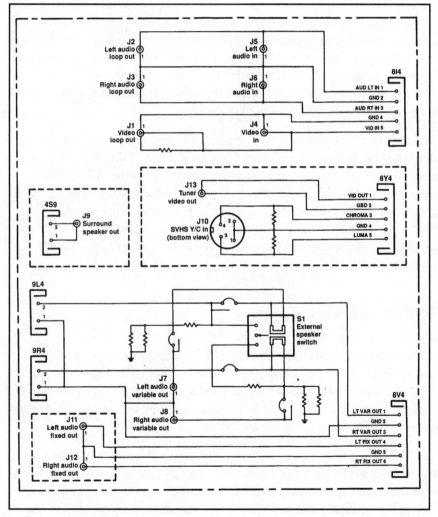

Figure 7-2. *By connecting the speakers to the "Variable Audio" jacks, I can control the speakers volume at my chair using the TV remote control.*

Wiring the Chair For Sound

I have had a recliner chair for years. The kids call it grandpa's rocker. The remotes are in the arm rests. From the chair I command a kingdom of consumer electronics products.

Pacing off about 20 feet from the TV, I cut 30 feet from a roll of wire (#18 bell wire) and fished the wire under the carpet from the TV to the chair. I positioned the speakers on the right and left top wings of the chair. Connecting the leads, I made certain the right speaker went to the right channel. The common ground went to both speakers. I phased both speakers using the old flashlight battery trick. When both speaker cones moved in the same direction with the same polarity I knew that they were in phase. With everything properly set up, I powered up the television, but not before I grabbed a glass of my favorite beverage and settled in the chair.

Forget about surround sound. This was even better. With my remote, I adjusted volume, balance and tone. I had no idea that TV stereo could be this good. I knew MTS stereo had only one-half the separation of FM radio, video disk or satellite TV, and much of what was there was lost in the small-screen TVs because the speakers are so close together. But this was great. What's nice about the "chair" is you have the same effect you get with earphones, plus the convenience of being able to hear the telephone or someone speaking to you.

Hearing Disability

A recent survey reveals that 11 percent of all people have some form of hearing disability. The percentage increases to 50 percent among people over 65. That equals a large market. Chair speakers are the perfect solution to this problem. You might be able to make some money by offering a service such as this to customers or potential customers.

Start by contacting potential customers. If any are interested, check to see if any of the chairs in the home are suitable for this type of treatment. Look to see if the set has controllable audio outputs. If all systems are go, quote them a price. With a high quote and your deposit check in hand, go back to the service center and assemble everything you need to make home installation easy.

I mounted tabs on my speakers that pass between the chair cushions to hold the speakers in place (*Figure 7-3*). The customer can adjust the exact speaker position to their preference. The speakers don't need to be in an enclosure for this application. The connecting wires will be out of sight as they are easily passed under the chair cushions.

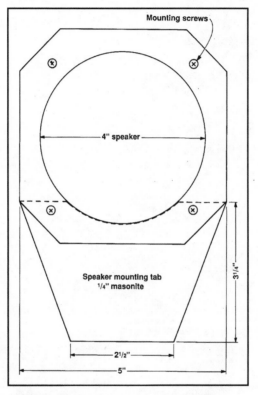

Figure 7-3. I attached tabs to the speakers as shown. With the tabs in place, mounting the speakers to the chair consisted of simply slipping the tabs between the cushions.

With a good survey and most of the assembly done in the service center, you can be in and out of the customer's home in less than an hour. If you want to upgrade the installation, add a circuit-closing stereo headphone jack for private listening. The output is a reasonable match for Walkman earphones.

Chapter 8

Restoring Scratched Compact Discs

By Matt J. McCullar

Compact discs, laserdiscs, and CD-ROMs take substantial abuse just from normal play. No matter how delicately they are handled by their owners, CDs pick up scratches and gouges from loading and unloading, and from spinning in the tray. Sometimes accidents happen; discs get dropped onto the floor and take a mean bounce or two, or they bang into the player tray when the user's hand slips.

Manufacturers claim discs will last indefinitely; what they fail to mention is discs last indefinitely only as long as they stay in the box. In the real world, their days are numbered just as with any other type of recording.

When the plastic is gouged or scratched, it may not be possible for the laser pickup to read the data from the disc below the gouge. When this happens, the CD player tries to fill in the missing data. If the scratch is severe enough, the listener hears a "skip."

While most stereophiles probably think that a disc with a scratch this severe is damaged beyond use and should be discarded, it is possible in some cases to smooth out the scratch so that the disc will play just as sweetly as it did when it had just been taken out of the box.

Even large gouges, given enough time and determination, can be eradicated so no skips remain. No longer will you or your customers have to toss out a disc because of a scratch. No longer must you purchase a new, expensive alignment test disc due to damage.

What Materials Do I Need?

The record-care product manufacturers sell a liquid that contains a fine abrasive that may be used to rub out scratches in CDs. There are alternative products that will also work, and are less expensive. Metal polishes, for example, contain extremely fine abrasive particles suspended in a liquid. Brasso, available at any hardware store, is very effective for rubbing out scratches in CDs.

You might want to experiment with other similar materials, but you should experiment on a disc that you don't care about until you're sure that the abrasive is fine enough that it won't cause scratches that will make the problem worse.

Also required are a few clean, lint-free cloths; the softer the better. Baby diapers work great. They will provide a soft surface to work on and will also polish away the scratch.

To repair a scratched disc, spread out a cloth on the workbench and place the disc on it, damaged side up. Shake the can of abrasive liquid and apply a little of it to another cloth. Then, holding on to the disc with one hand, use your other hand to wipe the polish onto the affected area.

Polishing out a scratch is very much like waxing a car. First you rub in a coat of the material, then you wait a while for it to dry, then you buff it off with another cloth. This takes some patience, because several minutes are required for the liquid to dry. But keep at it; it works. Make short, brisk strokes along the scratch, rather than across it. Replenish the polish with a fresh amount of liquid from time to time. Polish and buff.

Within a few minutes you will notice a difference. Hold the disc up to the light. You can see that the scratch is much less noticeable than before. It may take a lot of elbow grease, depending on how deep the scratch is, but it's well worth it. When you can no longer see the scratch, wash the disc with water and allow it to dry before attempting to play it.

As seen in *Figure 8-1*, polishing out a scratch involves polishing away plastic. The idea is to form a depression in the surface of the disc, the slope of which is gentle enough so the laser won't get confused when reading the data underneath it. With a scratch, the transistion from clear plastic to a large hole causes confusion.

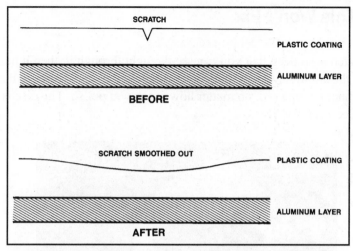

Figure 8-1. *Polishing out a scratch in a CD involves polishing away plastic. The idea is to form a depression in the surface of the disc, the slope of which is gentle enough so the laser won't get confused when reading the data underneath it.*

The polishing liquid won't damage a disc over the long term, but it is advisable to keep a fan going or open a window while you're doing this, to avoid prolonged breathing of the fumes.

Which Scratch is Which?

Today's players contain error-correction circuits that do a superb job in correcting most "data hits." But on a disc that contains many scratches, it may be difficult or even impossible to find out which one is actually causing the player to skip by just looking at it.

Since polishing a scratch can take some time, it is better to find out which one to work on in advance. The best way to do this is to listen to the disc and note how far into the music (or video, or data, or whatever) the errors begin. Keep it in mind that CDs begin at the center, and move outward.

A skip that begins within just a few minutes is likely to be located near the hub, and a skip that takes half an hour or more will be near the middle of the disc. A skip 60 minutes or more into the disc is located near the outer edge.

What This Won't Fix

If the scratch is so deep that you can see daylight through it, then it has damaged the aluminum layer that contains the digital data. Damage this severe cannot be repaired, no matter how much you polish. The disc is ruined. (See *Figure 8-2*)

Figure 8-2.

How Can I Use This?

I've used this technique on ordinary compact discs, laserdiscs, CDVs, CDIs, and CD-ROMs. Imagine how much it costs to replace a laserdisc or CD-ROM disc these days. Replacements have exorbitant prices, if replacements even exist. Knowing how to repair them is an excellent advantage, particularly if a disc used in the shop to align and test disc players suddenly reveals a wicked scratch.

You may try offering this service to your customers. How many times have you opened a CD player to find a disc stuck in it? It might be worth repairing a damaged disc for a customer. If you're a manager and don't want your technicians spending time on a project like this, assign it to someone else in the service center. It might work to your advantage.

Chapter 9

Servicing Audio Products

By John A. Ross

An audio-frequency, or A-F, amplifier is found at the last stage of a receiver sound circuit and amplifies only the narrow spectrum of audio frequencies that ranges from 20Hz to 20KHz. Audio output amplifiers should have the following characteristics: High gain; very little distortion within the audio frequency range; high input impedance; and low output impedance.

Modern audio and video receivers feature either transistor or integrated circuit amplifiers in the audio output stage that provide enough gain to drive a 30 percent modulated signal. Transistor-based audio amplifiers usually consist of a two-transistor, push-pull class B or AB output stage and, in small receivers, have an output range of 100mW to 1W. Audio amplifiers incorporated into an integrated circuit have an output power in the range of approximately 2W to 5W.

Audio amplifiers are also classified as power amplifiers. In an audio circuit, the power amplifier drives a high amount of power through a low resistance load with the speaker serving as the load. Although resistances dissipate power, efficient power amplifiers drive the maximum amount of power possible through a load. We can measure the figure of merit, or the efficiency, of an amplifier using the following equation:

n = (ac output power / dc input power) x 100%where n represents the figure of merit. Circuit designers rely on the figure of merit when matching an amplifier with an application in which it will be used.

Amplifier Classification

Any amplifier that conducts during the entire 360° of the ac input cycle is a class A amplifier. Class B and Class C amplifiers conduct for less than the entire ac input cycle. With a class B amplifier, two transistors conduct for

180° of the input cycle. One transistor conducts during the positive alternation of the ac input cycle and the other conducts during the negative alternation of the cycle. The combination of the conduction cycles yields a 360° output waveform.

A class AB amplifier is a variation of the class B amplifier. While the class AB amplifier also uses two transistors, conduction occurs only during the portion of the input cycle between 180° and 360°. Class C amplifiers use one transistor and conduct for less than 180° of the input cycle. Reactive components in the class C amplifier circuit provide the waveform for the remainder of the cycle.

In terms of efficiency, class A amplifiers have the lowest efficiency with a figure of merit ranging from 25 to 50%. The range depends on the use of RC or transformer coupled circuits. Class B and class AB amplifiers have an efficiency rating of approximately 78.5% while class C amplifiers have a maximum theoretical efficiency of 99%. Audio circuits rely on class B amplifiers because of the good efficiency rating. Although class C amplifiers have near-perfect efficiency, this type of amplifier works only with a tank circuit tuned to either the same frequency or a harmonic of the input signal.

A Class AB Amplifier

The amplifier circuit shown in *Figure 9-1* is configured as a class AB complementary-symmetry amplifier. When studying the circuit, note that one of the output transistors is an NPN transistor and the other is a PNP transistor. With this, two output transistors form a complementary pair and work like two variable resistors that are controlled by the audio signal amplitude. The NPN transistor draws current only during the positive half-wave of the audio signal and the PNP transistor draws current only during the negative half-wave. While complementary-symmetry circuits may use different components for biasing, the use of complementary transistors in the output stage will remain constant.

Transistors Q3 and Q4 provide biasing for the power amplifiers and act as diodes. With the biasing transistors configured in this way, each has a shorted collector-base junction and uses only the emitter-base junction in the circuit.

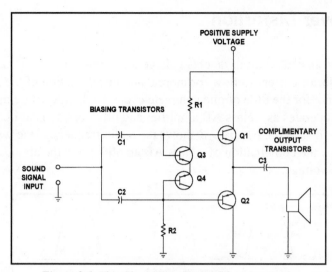

Figure 9-1. *This Class AB audio amplifier circuit uses complimentary output transistors.*

Using transistors in this way ensures that the biasing transistors match perfectly with the amplifier transistors. Because a complementary-symmetry amplifier does not require the use of a transformer, the circuit offers a low-cost, low-loss method for amplifying audio frequencies.

Because of the potential for crossover distortion when the two transistors go from conduction into cut-off, the biasing circuit prevents both transistors from entering cut-off during the same interval by maintaining a small amount of forward bias for each transistor. *Figure 9-2* shows the type of waveform produced by crossover distortion. Crossover distortion occurs when neither transistor in a class B amplifier has any forward bias and the output voltage of the amplifier circuit equals zero.

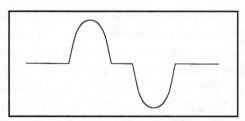

Figure 9-2. *Crossover distortion results in this type of waveform. This type of distortion occurs when neither transistor in a class B amplifier has any forward bias and the output voltage of the amplifier circuit equals zero.*

Crossover Distortion

Applying a small amount of dc bias voltage to the push-pull configuration allows collector current to flow for more than one alternation of the applied signal but not for the time of one complete cycle. Under these circumstances, the circuit operates as a class AB amplifier and produces a linear output that contains no distortion. *Figure 9-3* shows the audio frequency sine wave produced by the combination of a class AB amplifier and the application of the dc bias voltage.

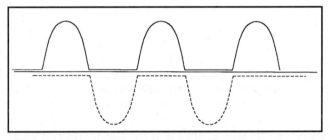

Figure 9-3. *This drawing shows the audio frequency sine wave produced by the combination of a class AB amplifier and the application of the dc bias voltage.*

The drawback caused by the need for added dc bias is the potential for thermal runaway. Often, power transistors are destroyed when an excessive forward bias voltage combines with junction leakage to produce progressively higher currents. With a fixed base current, the collector current increases. This higher-than-normal internal current causes the transistor to overheat which, in turn, breaks down the internal resistance of the device. As the cycle of increased heating and increased current production continues, the transistor eventually destroys itself.

As *Figure 9-3* has shown, one transistor of the class AB pair begins to conduct before the other has stopped conducting. For a brief time period, a path between power and ground exists through the transistors. To eliminate the thermal runaway problem, class AB amplifiers utilize a matched pair of transistors that have identical electrical and thermal characteristics. The use of a matched pair allows the same dc current to flow through both transistors and the same collector voltage to split equally between the two transistors.

In addition, the biasing circuit establishes a quasi-complementary-symmetry configuration. With this configuration, the complementary-symmetry section of the amplifier appears before the actual output stage. Rather than use a matched pair of high-cost output transistors, the circuit uses a matched pair of biasing transistors. Going back to *Figure 9-1*, the circuit consisting of transistors Q2 and Q3 provides the correct amount of forward bias for output transistors Q4 and Q5.

The circuit shown in *Figure 9-1* also provides an example of the amount of power dissipated in the resistances of a power amplifier circuit. Current flows through and voltage is applied across each resistor. Thus, using the power equation, or P = I2R, we can find the amount of power dissipated by each resistor. When we compare the sum of the dissipation amounts against the total power drawn by the amplifier, we have the total amount of power going to the load.

Every transistor has a maximum power dissipation rating. When selecting a replacement transistor for a power amplifier application, verify that the power dissipated by the transistor in the circuit does not exceed the rating of the replacement transistor. The PD(max), or maximum power dissipation rating of a transistor indicates the maximum amount of power in mW that a transistor can dissipate. For class B and class AB amplifiers, the maximum power dissipation rating equals:

$$P_{D(max)} = V^2_{PP} / 40R_L$$

where V_{PP} equals the peak-to-peak load voltage.

Variations In Class AB Circuit Amplifier Designs

Although *Figure 9-1* shows a popular class AB amplifier design, other types of class AB amplifiers also exist. Other popular designs include the diode-biased class AB amplifier, the Darlington complementary-symmetry amplifier, and the split-supply class AB amplifier. Each alternative offers characteristics that may be more useful for specific amplifier designs. *Figures 9-4A*, *9-4B*, and *9-4C* illustrate the different class AB amplifier circuits.

The diode-biased amplifier illustrated in *Figure 9-4A* uses two diodes to match the characteristic base-emitter voltage values of the output transistors. While class B amplifiers normally operate at cutoff, the addition of the diodes biases the transistors above cutoff. With both amplifiers operating above cutoff and conducting between 180° and 360° of the waveform, the circuit has the response shown in *Figure 9-3* and begins to operate in the class AB range. Thus, because of the matched characteristics, this variation of the circuit eliminates the chance for thermal runaway or crossover distortion.

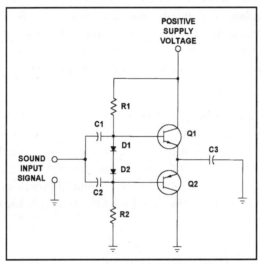

Figure 9-4A. The diode-based amplifier uses two diodes to match the characteristic base-emitter voltage values of the output transistors.

Darlington Pairs

In the circuit of *Figure 9-4B*, two Darlington pairs function as the output stage and provide higher overall current gain than the standard complementary-symmetry amplifier. Although both transistors of a Darlington pair are usually packaged in the same case, the term actually describes the configuration of the two output transistors. The first transistor of the pair acts as an input amplifier for the second transistor. Because the dc and ac beta values of the pair equals the product of the individual transistor betas, the Darlington pair has an extremely high beta. With the emitter of the first transistor connected to the base of the second, the base current of the first is multiplied by the beta of the second.

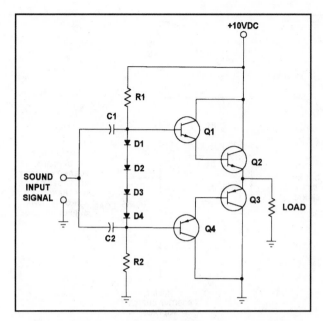

Figure 9-4B. *Two Darlington pairs function as the output stage and provide higher overall current gain than does the standard complementary-symmetry amplifer.*

Usually, a Darlington complementary-symmetry amplifier is used for applications that require high load power. Because of the use of an input amplifier, which the Darlington configuration provides, the transistor pair has better stability, high current gain, and a high input impedance. The four diodes shown in the diagram provide biasing at the bases of Q1 and Q3 and compensate for the base-emitter voltage required for each Darlington pair.

In the split-supply class AB amplifier shown in *Figure 9-4C*, the two power supply connections have equal but opposite polarities. By using matched power supplies, each amplifier drops its own supply voltage. This allows the output of the class AB amplifier to center around zero volts rather than the division of the supply voltage.

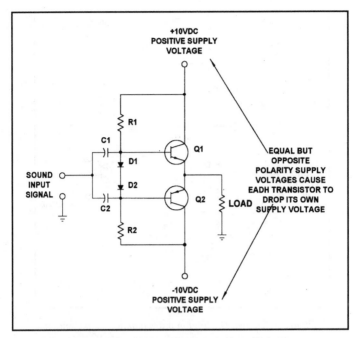

Figure 9-4C. The voltages at the two power supply connections in the split-supply class AB amplifier have equal but opposite polarities.

Integrated Circuit Audio Amplifiers

Figure 9-5 shows the schematic diagram for an audio amplifier found within a single IC package and based on an operational amplifier. The op-amp provides a combination of high input impedance, low output impedance, and low noise distortion. With the low output impedance and low noise distortion, the amplifier provides optimum coupling to the speaker with minimum distortion in the audio frequency range.

In the circuit, the grounding of the operational amplifier negative supply voltage input through capacitor C5, a coupling capacitor, limits the amplifier output to a specific range. Grounding the supply in this way places the reference for the speaker close to ground, and, because of its location in the bias supply voltage line, protects transistor Q1 from transient current. Without the capacitor in that location, current could feed back from the operational amplifier through the power supply and to the transistor.

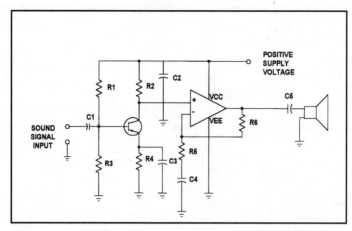

Figure 9-5. This is the schematic diagram for an audio amplifier found within a single IC package and based on an operational amplifier. The operational amplifier provides a combination of high input impedance, low output impedance, and low noise distortion.

Newer sound channel designs enclose the complete audio system into one or two integrated circuits. *Figure 9-6* shows the schematic/block diagram for the sound processing system for a Sylvania model 19C5 television. In the figure, IC111 handles the amplification of the 4.5MHz sound i.f. signal, provides limiting, detects the FM audio signal, and controls the level of the signal before coupling it to IC121, the audio output IC.

As the figure shows, the composite video signal travels from pin 12 of IC51, the video i.f. processing IC, through a coupling capacitor and to a 4.5MHz high-pass filter. From there, the 4.5MHz sound i.f. signal travels to a 4.5MHz buffer transistor and then to T110, a sound input transformer. After arriving at pin 14 of IC111, the 4.5MHz is amplified, limited, detected, and coupled to the IC121. The +12Vdc voltage supply seen at pin 11 of IC1 and the Q36 collector circuit has its source at the integrated flyback transformer.

Moving to pin 5 of the sound i.f. IC, a voltage divider controls the volume. R114, the audio preset control, compensates for any variations within the sound i.f. IC by controlling the audio level when the volume control is set to its minimum level. The controlled audio signal couples from pin eight of IC111 through a 1(F capacitor to pin nine of IC121.

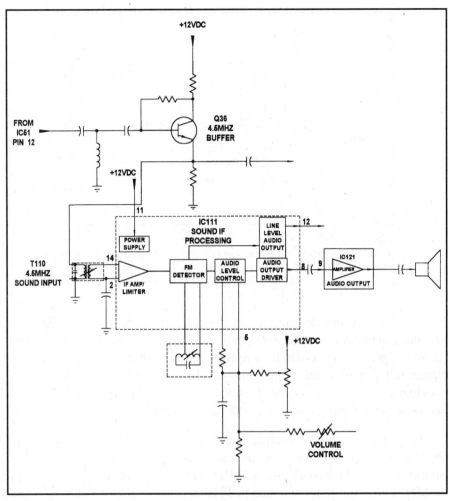

Figure 9-6. This is the schematic/block diagram for the sound processing system of a Sylvania model 19C5 television. IC111 handles the amplification of the 4.5 MHz sound i.f. signal, provides limiting, detects the FM audio signal, and controls the level of the signal before coupling it to IC121, the audio output IC.

Troubleshooting Audio Circuits

If we consider the audio stage as something that provides the power to mechanically move a speaker cone, then troubleshooting the stage may seem easier. With audio frequencies, the stage generates power throughout a range from 20Hz to 20kHz. Any time that a stage provides power, it also becomes exposed to the possibility of a short or an overload. Either condition can cause symptoms such as blown fuses, open resistors, and shorted power transistors to surface.

As a rule, always look for one of these three basic problems when trouble-shooting audio circuits:

1. The bias current may have risen to a high level and caused a higher-than-normal voltage drop across the emitter resistors in a transistor-based ampli-fier circuit. Under normal operating conditions, the voltage drop will stay within the 10mV to 30mV range. With a high bias current, the voltage drop may measure as high as 10V. Too much bias current will cause the amplifier to heat while too little bias current will cause distortion at low volumes.

2. Consider high-frequency oscillation as a possible problem. If high-fre-quency oscillation exists, the amplifier will get unusually hot with no applied signal and will have a low emitter voltage. More than likely, high-frequency oscillation will also cause hum and decrease the amount of power at the speaker.

3. Finally, clipping may occur at different signal levels and may be seen at the top and bottom of the output waveform.

As with all troubleshooting efforts, working with the audio section requires a logical approach and basic knowledge about the circuit operation. For example, if the speakers, fuses, and connections appear in working order, measure the power supply voltage. At the pre-amplifier stage, filters mini-mize hum and noise while providing a very stable supply voltage. DC voltages at the driver and power amplifier stages connect directly to capaci-tors that smooth any ripple.

After verifying the presence of the correct supply voltage, use common signal injection and tracing techniques to determine where the fault exists. With the audio stages, we can inject an audio frequency at the beginning of the stage and check for the presence of a signal at the output of the stage. Usually, the schematic will indicate a good point for injecting the substitute signal.

Applying Test Equipment To the Problem

Troubleshooting audio amplifier problems requires four basic pieces of test equipment. An isolation transformer will protect you, your test equipment, and the equipment under test. The use of a variable transformer will allow the testing of an amplifier under different voltage conditions and will allow the gradual application of power. An oscilloscope, multimeter, and an audio frequency generator provide the equipment necessary for checking both the input and output signals, component and stage voltages; setting the bias, and for finding high frequency oscillations.

When injecting a signal into an audio channel, use an audio or function generator to supply either a 3kHz square or sine wave. In addition, connect either an 8Ω or a 10Ω resistor across the stereo speaker terminals as a load. The wattage rating of the resistor chosen as a load should match the power output of the amplifier. For transistor amplifiers, connect the audio or function generator to the first audio transistor. Integrated circuit audio channels will require a connection to the pre-amp portion of the circuit.

Then, clip a test lead from the generator probe to the input channel. By using an oscilloscope, you can monitor the performance of the audio channel while injecting the test signal. Set the variable sweep control of the oscilloscope for a steady waveform and adjust the input signal so that a stable waveform appears. When injecting a sine wave into the circuit, the shape of the sine wave should not change as the signal progresses from the input to the output of the amplifier.

Squarewave Response

Depending on the shape and amplitude of the waveform at the output of each stage, the injection of a square wave signal can tell you about the type of problem occurring within the circuit and disclose the source of the problem. For example, a square wave with a rounded leading edge indicates the loss of high frequency response in the amplifier stage. Using a square wave as an input signal also allows the technician to test for internal oscillation, or ringing. Ringing may occur any time that a defective component, such as an output transformer, introduces oscillation into an amplifier circuit. *Figure 9-7* shows several square waves and the corresponding problem as indicated by the shape of the waveform.

Using a square wave input signal allows a technician to verify that the amplifier passes all the harmonic components of the fundamental frequency and that it has sufficient bandwidth. When injecting and monitoring a square wave, check the amplifier response at a range of low and high frequencies. As *Figure 9-7* shows, the waveform will appear differently for poor low frequency or poor high frequency response conditions. For example, the waveform at an amplifier output for an injected low-frequency signal may show a perfect square wave while the waveform for an injected high frequency signal may show some rounding at the edges.

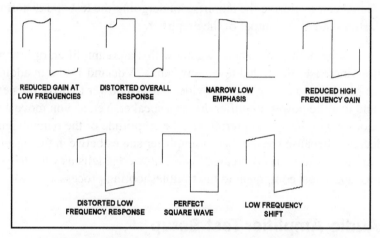

Figure 9-7. Using a square wave substitute input signal allows a technician to verify that the amplifier passes all the harmonic components of the fundamental frequency of the waveform and that it has sufficient bandwidth.

However, some rounding of the square wave is permissible at the higher frequencies. Even when the amplifier is not up-to-standard, it may continue to produce excellent sound. This occurs because response is measured relative to a repetitive frequency. An amplifier would need an extremely wide frequency response of 10,000Hz to 200,000Hz to produce a perfect square wave. On the hand, a severely-rounded square wave, which almost resembles a sine wave, indicates that the circuit has excessive high frequency attenuation and that the amplifier has eliminated most of the harmonics associated with the fundamental frequency.

Monitor The Amplitude

As with many of our tests, we can check the overall response of the circuit by monitoring the amplitude of the injected signal from the output of the circuit back to the first input stage. For example, a known-bad circuit may have a strong output signal with the injection occurring at the volume control. However, moving the injection point further away from the output may disclose that the next stage has a defect. When injecting the signal, remember to keep the signal generator input signal to a low level. An overly-strong input signal can overdrive the amplifier undergoing the test and distort the waveform found at the output of the amplifier.

Even if no schematic is available, we know that a preamplifier, or low-noise amplifier, increases the signal level between 0.7Vdc and 1Vdc. In addition, we also know that a driver stage provides enough power to drive the next stage and that the power amplifier drives a speaker. Yet, as you move a signal generator toward the speaker terminals, the amplitude of the output signal will decrease because the number of amplifier stages between the injection point and the speaker has decreased. This knowledge tells us what to expect when we apply test equipment to the troubleshooting process.

An Audio Amplifier Test Setup

Figure 9-8 shows a typical test set-up needed for checking the performance of an audio amplifier. Moving from left to right, an audio frequency generator that has a frequency range of 10Hz to 20kHz works as a signal source. The generator supplies an adjustable output voltage that ranges between 0.5mV and 2V. While the oscilloscope provides a method for monitoring the output signal, we can measure the output voltage with a multimeter.

In addition, we can use the oscilloscope to check for noise in the audio channel. Checking the voltages at the pre-amplifier with an oscilloscope can disclose the source of the noise. In many instances, active components in the signal path can develop characteristics that generate noise. In others, an overloaded amplifier stage can produce distortion.

Whenever an output transistor or integrated circuit becomes leaky, the problem condition will introduce distortion into the signal. Injecting a signal into the audio channel stage-by-stage should allow you to narrow the troubleshooting process. At this point, you can use your knowledge about transistors or integrated circuits to find the faulty stage. Distortion at the output but not the input of either a transistor or integrated circuit audio channel points towards that stage.

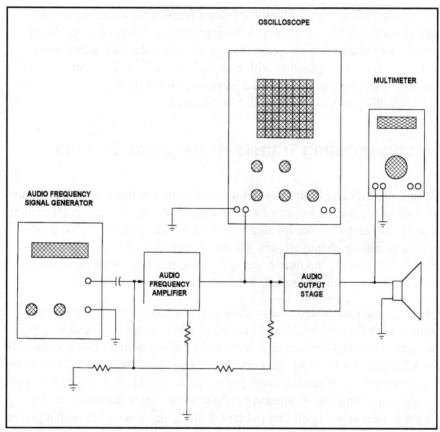

Figure 9-8. *A test set-up such as this allows the technician to check the performance of an audio amplifier. The audio frequency generator has a frequency range of 10Hz to 20kHz and supplies an adjustable output voltage that ranges between 0.5mV and 2V. The oscilloscope provides a method for monitoring the output signal. The mulitimeter provides for measurement of the output voltage at the load resistor. In addition, the oscilloscope can be used to check for noise in the audio channel.*

If the amplifier stages utilize capacitor coupling, always check the output capacitors. Many times, a dried electrolytic capacitor in the output stage will either badly distort the signal or cut the amount of power delivered to the speaker. Under normal circumstances, an increase in frequency will cause the capacitive reactance of a capacitor to decrease and the circuit current to increase. When an electrolytic capacitor dries, the internal resistance of the capacitor increases, dissipates additional power, and cuts the ability of the circuit to reproduce high frequencies.

One quick method for checking for a dried electrolytic capacitor involves measuring the voltage drop across the capacitor. In a normally-operating circuit, the voltage drop decreases to a negligible value as the frequency increases. A circuit operating with a defective coupling capacitor will have a larger voltage drop. Because of the increased power dissipation within the capacitor, the part may also feel hot to the touch.

Troubleshooting Transistor Amplifier Circuits

Class B and AB amplifiers pose some interesting troubleshooting problems because of the use of two transistors in the output and biasing stages. Either of the two transistors can develop a shorted or open junction or have inter- mittent operating characteristics. The most common fault associated with amplifier circuits is "no output signal." Often, this type of problem leads technicians on a fruitless search for a defective output transistor.

Often, simple time-tested checks such as signal injection and tracing will point toward the problem. More than likely, though, we will need to apply our knowledge of amplifier circuits to our troubleshooting efforts. Class AB amplifiers draw very little idle current. As a signal is applied, the amount of current increases. When the amplifier is operated in class A mode, the circuit draws a small amount of quiescent current to minimize distortion at low output power. When operating in class B mode and at a higher output power, the amount of current drawn from the power supply depends on the amount of output power.

In addition, each active component of a class AB amplifier amplifies only a half wave of the audio signal. The half-waves of the output signal combine at the output stage of the amplifier. Both of these factors affect what we might see if we monitor the audio stages with an oscilloscope or multimeter.

Along with these factors, we should also remember that audio output stages usually do not rely on an output transformer. Some transformerless power stages will have positive and negative supply voltages while others will rely on a "center voltage" for both power transistors. The center voltage at each transistor always equals half of the supply voltage. In all situations, because a dc voltage will harm a speaker cone, you should see 0Vdc at the output. From a troubleshooting perspective, we can begin to look for a very high dc voltage at the output as one fault indicator or a very low or negative voltage at the positive supply as another.

Check the Sound Channel

Before focusing on only the output stages, check the sound channel for the correct input signal from the previous stages, the correct supply voltages, and good ground connections. After confirming the presence of the signal, voltage, and ground connections, disconnect the load from the circuit and recheck the voltage and signal characteristics. At times, the load can prevent an amplifier from working by adding additional loading. In this case, you would check for a shorted load.

If all these circuit checks point to the amplifier, disconnect the ac signal source, or the output of the previous stage, from the output stage. To do this, either desolder a connecting wire or remove the coupling component that connects the two stages. Either method will isolate the output stage. After isolating the output stage, check the voltages at the base, emitter, and collector terminals of each transistor. Also, check for an even distribution of voltages at the voltage dividers. *Figure 9-9* illustrates the voltage checkpoints.

A leaky transistor will have lower-than-normal voltages at both the collector and base. Along with those checks, verify that the transistor has the correct forward bias between the base and emitter terminals. A leaky transistor will have an improper forward bias. Many times, though, the base or emitter bias resistors can change value and cause the amplifier to appear leaky.

An overload or shorted output will cause a power transistor to short. To find which power transistor has shorted, check the resistance between the collector and emitter of each transistor. Furthermore, check the corresponding driver transistor. Many times, a breakdown within the emitter-base or collector-base junctions of the power transistor will cause both the output transistor and the driver transistor to overload. Generally, the driver transistor will have an emitter-base short. As a final rule-of-thumb, if the symptoms direct your efforts towards one amplifier transistor, replace both transistors. More than likely, a defect in one transistor has damaged the other.

Each of the voltage checks tells us about the current operating conditions of the circuit and allows us to pinpoint the problem area. Table 1 lists some of the common symptoms and faults found with transistor AB amplifiers. The component numbers listed in the table correspond with the numbers listed in *Figure 9-9*. Normal voltage readings for the table are taken from the schematic drawing for the particular circuit.

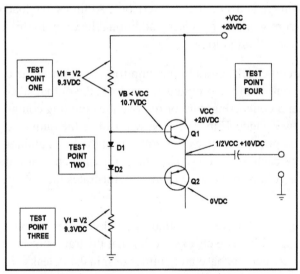

Figure 9-9. If you have isolated the cause of an audio problem to the output stage, check the voltages at the base, emitter, and collector terminals of each transistor. Also, check for an even distribution of voltages at the voltage dividers. The check-points are shown here.

Troubleshooting Integrated Circuit Amplifiers

Because of the level of integration seen in IC-based sound circuits, the search for a problem solution is limited to only a few points. The application of basic signal tracing skills and voltage checks will often disclose the source of the problem. As *Figure 9-10* shows, we can apply our knowledge about feedback and gain while checking the circuit from output to input. In the figure, R1 is in a negative feedback path. The combination of R1 and R2 determine the gain of the power amplifier while resistor R3 stabilizes the output voltage.

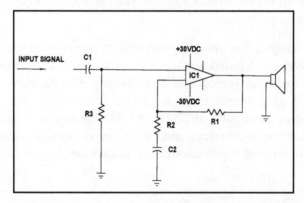

Figure 9-10. *Because of the level of integration seen in IC-based sound circuits, the search for a problem solution is limited to only a few points. The application of basic signal tracing skills and voltage checks will often disclose the source of the problem. A knowledge about feedback and gain helps.*

First, with a signal applied to the receiver, check for an audio signal at the output of the audio amplifier. Next, check for the correct voltage at the source voltage connection of the IC. If those two checks provide no hint for a solution, move your efforts to the IC terminals that connect to the volume control.

Checking Stereo Amplifiers

One of the nice things about troubleshooting stereo amplifiers is that every stage has two identical signal channels. If one channel fails, the other channel works as a reference for verifying the presence of dc voltages and audio frequency signals. For this reason, we can use a dual-channel oscilloscope to compare the performance of the audio channels with each other. So that you can maintain accurate measurements, adjust the amplifier balance control for even balance between the two channels and adjust the oscilloscope gain controls for each channel to the same levels. In addition, adjust the variable sweep control of the oscilloscope so that you can see two stable waveforms on the oscilloscope display.

At times, a change in the frequency response of one channel will cause that channel to produce distorted, low-quality sound. Using an oscilloscope, compare the frequency response of both channels with the curve shown in *Figure 9-11*. A good amplifier will have a flat frequency response for frequencies ranging from 20Hz to 20kHz. As with the monaural sound channel, we can also use the oscilloscope and a signal generator to monitor the progress of an injected signal through an audio channel.

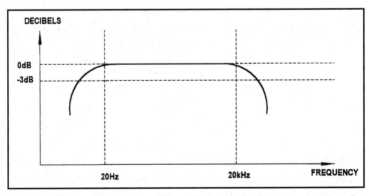

Figure 9-11. A good audio amplifier will have a flat frequency response for frequencies ranging from 20Hz to 20KHz.

When the frequency response of one channel decreases, inject a 400Hz square wave into the input of the audio stage and use an oscilloscope to check the shape of the signal at the stage output. Checking the frequency response then involves setting the generator to reference value 1000Hz and the output signal to a value below the maximum output power. At this point, the volume control should be set to the minimum level.

Switching the signal generator frequency from 20Hz to 20kHz should not cause the output voltage to change by more than 1dB. As mentioned, the frequency response should remain flat. The amplifier should begin to symmetrically clip the positive and negative peaks of a sinewave at the output as the input voltage increases. The key phrase here is "symmetrically clip." If the oscilloscope shows less-than-symmetrical clipping, one amplifier does not have the proper frequency response.

After checking the response of the distorted channel, begin to narrow your search by checking for bad electrolytic capacitors in the amplifier line. As mentioned, a bad coupling capacitor will harm the frequency response of an amplifier stage. When using signal injection in an effort to find the bad component, bypass both the pre-amplifier and the tone control, and consider the point where the signal begins to clip. An amplifier that has poor frequency response will have an output signal that begins to clip at a low output power setting.

Summary

The two examples shown in this summary are basic applications of the troubleshooting skills described in this article. Solving the problems shown in the examples becomes a matter of employing logical troubleshooting techniques and applying test equipment to the symptoms. Some of the techniques include careful observation, waveform comparison, voltage and resistance measurements, and component substitution. In addition, troubleshooting the audio problems requires knowledge about how the amplifiers operate and how components in the amplifier stages can affect that operation.

Chapter 10

Servicing Musical Instrument Electronics
By Ron C. Johnson

In past articles I've suggested a number of ways you and your shop could use expertise you already have to diversify into new areas. Your test setup, personnel and skills are transferable from one service area of electronics to another. Moving into a new area really isn't all that difficult. The trick is to find an area that's profitable. One of the goals of this column is to help you do that.

Since traditional consumer electronic servicing includes audio equipment such as tuners, receivers and power amplifiers, servicing musical instrument electronics is just a small step away. In this issue I'd like to talk about some technical and some non-technical aspects of musical instrument electronic servicing that may be of help to you.

A List of the Equipment

In case you're not the musical type and are not familiar with the kind of equipment you'll find in a music store, here's a short list.

Musical instrument amplifiers for guitars, basses, synthesizers, etc. are the main items to be found here. Guitar amps usually have a built in speaker system and the electronics are all over the map: transistor amps, linear power blocks, and even tube amps which deliver that special sound (some would say distortion) desired by musicians.

Another big area is sound reinforcement systems. Sound systems are often rented to bands. A typical sound system consists of sound mixers, equalizers, and effects (digital delays, reverbs, etc.) as well as heavy duty power amps and large speakers. System rentals take a lot of abuse from moving and from rough usage which can provide you with ongoing service work.

Synthesizers of all shapes and descriptions require service from time to time. Again, these are sometimes rented, and therefore often require repair. The level of technology in synthesizers is pretty high and can be a real challenge. MIDI, a serial communication system between synthesizers, is the heart of the latest equipment. You may have come across MIDI in personal computers that have multimedia options. Building or repairing custom MIDI interfaces and cables can add to your service income.

Electric guitars are being manufactured with an assortment of specialty pickups, active electronics and synthesizer interfaces. There is quite a bit of special knowledge required here but it can be found in a few good books on the subject. Mechanical repairs to the guitars (and other instruments) can provide some work too, if you have the expertise to do it.

Fixing the Cables

Cable repairs can keep you busy as well. Sound systems use "snakes," multiconductor cables for running multiple microphone lines from the mixing board to the stage. These get beat up and need work regularly, especially the connectors on each end. You can also build and sell them in your spare time. Building snakes isn't difficult but you need to keep costs down to be competitive.

Most music stores also sell and rent recording equipment, usually special multitrack recorders. Some of these recorders use cassettes, others use video tape cartridges, and some are reel to reel. Your expertise in VCR's and audio cassettes can be useful here.

Business Considerations

Before we go on to talk about some of the technical aspects of doing this kind of work, some other business considerations are important. First, if the store is large, or if it is part of a chain, it may have its own service center and service personnel. The ones that don't, probably contract the service to a company such as yours, or they have somebody who comes in part time to do it.

If you find a music store that needs someone to perform their service, you'll have to decide whether to send one of your technicians to work at the music store or have somebody haul all the repairs back to your service center. This is no trivial matter. Often the music store wants service done at its location because some of this stuff is big and heavy.

We all know that it's difficult to make service calls profitable and this is no exception. Travel time to and from the store, acquiring parts, down time waiting for store personnel and other delays can erode your profit margin.

You need to get some questions settled at the outset of your relationship with the music store. For example, can you charge for checking out equipment that proves out as being good? It's important that you are organized and have a clear understanding with the music store.

If you have a work area in the store, is it yours exclusively? Some stores do speaker reconing and guitar adjustments but no electronics. When I was doing this kind of work I often came in on my assigned day to find the work area a mess. The store didn't like me charging them to clean it up but somebody had to do it. The final straw came when I found a pair of my needle nose pliers stuck to the workbench with spilled speaker cement all over them. The store and I parted company not long afterwards.

Technical Considerations

Let's zero in on the technical end of musical instrument repairs and see some of the similarities and differences between what you're doing now and what you're likely to find.

Musical instrument (MI) power amplifiers have some similarities to the ones you see regularly. The heart of a consumer stereo system is the power amp that takes signals from a preamp or receiver and drives the speaker system. The same is true of MI amplifiers. Guitar amps, for instance, usually have two channels of preamplification and an output driver. Sound system amps usually have one or two line level inputs to the power amp section.

While a typical consumer stereo amp is generally designed to be mass produced, MI amps have the added requirement of being ruggedly built to hold up under hauling them around, dropping them and driving them beyond their design limits. A number of other features further differentiate them from consumer amps. For instance, MI amps have built in reverb circuits, tremolo, special filters, compressors and other effects. For now, let's take a look at the power amplifier section itself.

The Musical Instrument Power Amplifier

Some of the amplifiers you repair now may be capable of fairly high power output, but chances are most of the stuff you see is relatively low power and uses hybrid power ICs in the final stage. Hybrid power ICs are used in some of the smaller practice amps and a few of the larger ones, but by and large you'll still find a lot of the good old Class AB push-push amplifiers. A few of the newer amplifiers use VMOS FETs in their outputs, but, in my experience, being a bit unstable, they have never really caught on. They tend to fail catastrophically, taking out several components at once.

You might be surprised at how many guitar amps still use tubes. Some of them use tubes throughout while others only use them in one or two stages. There's something about tube amps that musicians like. Although tube circuits do create some harmonic distortion, they cause less intermodulation distortion than solid state circuits. The result is a smoother sound, or so say the musicians.

A Real-world MI Amplifier

Figure 10-1 shows the output stage of a typical musical instrument amplifier. This one is out of a Fender guitar amp, has been around for a few years, and hasn't changed much. I'll use it to point out some of the more common problems with this kind of amp.

This is a direct coupled Class AB amplifier that puts out about 40W RMS. Notice the final output transistors, Q202 and Q203 are both NPN with 0.5Ω, 5W resistors between them. This is called a quasi-complementary symmetry amplifier. If an amp like this fails, usually one or both of the output transistors will short out.

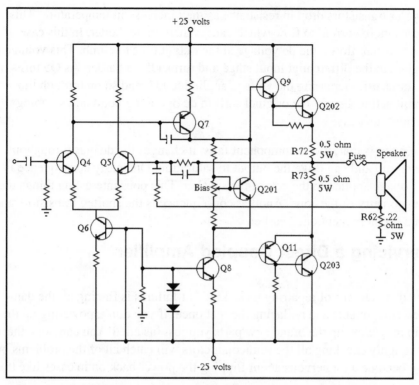

Figure 10-1. The output stage of a typical musical instrument amplifier. This one, out of a Fender guitar amp, has been around for a few years, and hasn't changed much.

If you encounter an amp like this one that's not working, always check the two resistors, R72 and R73, as they will often open up, and usually there is no visual indication that they carried excessive current. Another resistor to keep an eye on is R62, the 0.22Ω, 5W resistor that connects the speaker to ground.

Now take a look at the differential input stage to the amp (Q4 and Q5). The input signal is applied to the base of Q4 and a negative feedback signal is applied to Q5. The feedback signal comes from the junction of R72 and R73, which is also the actual amplifier output to the speaker. The dc level at this point should be zero volts; the output current to the speaker should swing positive and negative around this point. The feedback voltage from this point reduces the overall gain of the circuit and minimizes distortion. It also serves to work against thermal runaway in the output stage.

Bipolar transistors drop in resistance as they increase in temperature. This causes more current to flow which causes them to get hotter. In this case, if more current flows, the dc voltage at the output tries to climb. This voltage is applied to the differential input stage and turns Q5 on harder. As Q5 turns on Q4 turns off, increasing the voltage applied to Q7, and so on through the circuit acting against the original shift in dc operating point and stopping thermal runaway.

The point is that, if one component fails, it changes the dc bias throughout the circuit and can drive the output to saturation. One nasty consequence is dc voltage applied to the coil of the speaker. This does interesting things to the geometry of the cone. Another consequence is the shorted transistors and open resistors mentioned earlier.

Servicing a Direct Coupled Amplifier

The difficult part of repairing these kinds of failures is finding all the damaged components and replacing them at once. If you don't, powering up the amp may blow up the brand new parts you just installed. You can hope that thoroughly checking all the semiconductors will catch all of the problems but I've become a bit nervous about flipping the power back on in cases like this.

One method of testing to see if a fault still exists is to make up a power cord with a low wattage light bulb in series with the hot conductor. When you turn on the power, if the amp is still faulty and tries to draw lots of current, the bulb will light up brightly. The initial high resistance of the bulb may limit the inrush current enough to save the parts you have installed, if you turn off the power quickly enough—maybe—hopefully.

A better way to check your work is to use the setup shown in *Figure 10-2*. An isolation transformer feeds a variable transformer and a special ac power cord (you make up) that allows you to monitor ac current. You bring up the ac voltage slowly, watching the ac current. Most of these amps will draw less than 1Aac with no signal applied. If the current keeps on rising through about 1A as you increase the voltage something is probably still faulty in the circuit. Shut it down and start over.

By the way, the load shown is a dummy speaker load made up of two 200W, 8Ω resistors mounted on standoffs on a board. This is useful once you get the amp working to check for crossover distortion and symmetrical clipping at full output.

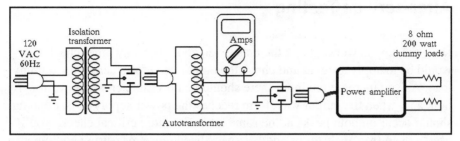

Figure 10-2. *Use this set-up to apply power to a direct coupled amp once you have completed servicing it. An isolation transformer feeds a variable transformer, and a special ac power cord (you make up) allows you to monitor ac current. You bring up the ac voltage slowly, watching the ac current.*

The unfortunate part of troubleshooting a circuit like this is that you seldom get a small problem with it. If one component changes the dc bias, everything goes out of line. Even a shorted capacitor will cause this kind of problem.

The Finishing Touches

Once the amp is working, the only other consideration is the bias adjustment. The bias pot is what makes the amp Class AB instead of Class B. Its job is to bias Q201 on so that a fixed voltage is applied from the base of Q9 to the base of Q11. This voltage will just begin to turn on those transistors, as well as the output drivers, Q202 and Q203.

With a small base current flowing already, any ac signal applied will be amplified linearly. Without this, crossover distortion would result. Q8 is a current source (its emitter current is fixed by the two diodes from its base to the negative rail). Keeping the current through the bias circuit constant stops signal fluctuations from changing the bias voltage.

There are a couple of ways to adjust the bias pot. One crude method is, with an input signal driving the amp, monitor the signal across the dummy load and adjust the pot for no visible crossover distortion. I'm sure you can guess how accurate that might be. A better way would be, with no input signal or load, to measure the voltage across the emitter resistor, R73, and adjust for about 12mV (about 25mA emitter current). Better yet, use a distortion analyzer and adjust for the specification given for the amplifier.

After-service Testing

Once the amp is working and the bias is set, about the only other check would be to apply a signal and check the output waveform across the dummy load. As you increase the input there should be no visible crossover distortion, and, when the output waveform reaches the power supply rails, clipping should occur on both peaks at the same time. For more critical checks you'll need a distortion analyzer or another specialized noise measuring equipment.

There is a real variety of interesting technology to work on in the musical instrument field and money to be made doing it. In addition to rental repairs and repairs to customers' equipment, warranty repairs are available. By making contact with equipment manufacturers through the music store you can become the authorized warranty center for the brands handled by that store and eventually widen your market to include others. As usual, good business sense and constant attention to the bottom line are just as important as technical ability.

In a future installment, I'll take a look at some other aspects of musical instrument servicing with some more tips on the technical end of the job.

Chapter 11

Should You Consider Commercial Sound?— Public Address, Paging and Intercom Systems

By Ron Johnson

What's the difference between typical consumer audio equipment and the commercial sound equipment used in public address, paging and intercom systems? That's what I'll be talking about this month. If you're interested in servicing this kind of equipment, or selling and installing it, there are a few important points you should know.

Public address systems can be divided into various categories, in various ways. One way is to split them down the middle into lower-end commercial public address systems, and high-end professional sound reinforcement systems.

The low-end, or commercial types of systems, can be further divided into paging systems, intercoms, background music systems, distributed sound, fast-food drive-through systems, and a number of variations and combinations of these.

The high-end systems are very similar to the professional sound reinforcement systems used by touring musicians, except that the equipment is permanently installed. If you work further into this area you can get into night club, recording studio, and broadcast equipment.

Since the "bread and butter" of sound contracting is commercial equipment let's take a look at some of the features of a typical medium-power public address system.

The imaginary system we're going to talk about is a sound system for a small grocery store. It will provide background music from a consumer tape deck or AM/ FM radio which will be muted when the press-to-talk switch on a paging microphone is depressed. A commercial music source could also be used if the customer desired. The microphone signal is connected to the heart of the system, a commercial sound distribution amplifier with a 70V output, which drives inexpensive ceiling speakers. This particular amp, the Rotel QA-100, one I used frequently when I was in this business, also has a built in AM\FM radio.

Figure 11-1 shows a block diagram of the components of the system. As we discuss each section of the system I'll point out some of the features that make this equipment different from other equipment you may have worked on.

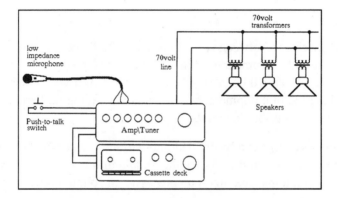

Figure 11-1. *A simple commercial sound system might look like this.*

Microphones

The system in *Figure 11-1* shows one microphone, with a three-conductor cable, connected to the amplifier. This amplifier can accept two low-imped-ance microphones (balanced or unbalanced) or one low-impedance and one high-impedance microphone. The low-impedance mics are connected to screw terminals on the back of the amp, while the high-impedance one uses a quarter-inch phone plug, which plugs into the front panel.

If you're new to microphones or (like the rest of us) need to brush up on your theory about input impedances, etc., the next couple of paragraphs will give you a quick overview.

Microphone Impedance

I've already mentioned that there are two categories of microphone impedances: low and high. Although there is quite a bit of theory behind the
subject, for practical purposes we can simplify this a bit.

First, low-impedance microphones are used when running long microphone
lines (up to about 200 feet) because they don't attenuate the high frequencies
as much as high-impedance microphones do. Think of the microphone itself
as an ac voltage source with an internal resistance (or impedance) in series
with it. The cable looks like a parallel capacitor which increases in value
with length (*Figure 11-2*).

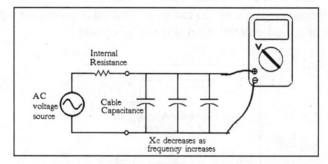

*Figure 11-2. In a commercial sound system, the microphone cable
appears to the microphone like a parallel capacitor that increases in
value with length. As the frequency of the signal increases, the
impedance of the capacitance will drop.*

As the frequency of the signal increases, the impedance of the capacitance
will drop. The microphone internal impedance and the capacitive reactance
of the cable form a voltage divider in which the voltage across the capacitive
reactance is the level of the signal reaching the amplifier. As the signal
frequency increases, the signal reaching the amplifier decreases. The higher
the microphone impedance the worse the problem becomes because voltage
is dropped across the mic impedance and never reaches the amplifier.

All of this just indicates that lower mic impedances allow you to run the
wiring farther without degrading the high-frequency response of the signal.
For short distances, high-impedance mics work fairly well. The advantage of
low-impedance microphones is offset somewhat by the fact that their signal

levels are lower and, because of this and the longer lines involved, induced noise can become a problem. To overcome the problem, balanced lines are used.

Balanced vs. Unbalanced Lines

Figure 11-3 shows the wiring for balanced and unbalanced lines. Balanced lines use two conductors inside an overall shield. The mic output is connected to the two lines and the shield is grounded to the amplifier chassis (which should be at earth ground). Grounding the shield causes induced noise currents to be shunted to ground. You can think of the shield as a magnetic barrier around the signal lines. In addition, any noise induced onto both wires (common-mode noise) can be subtracted out using an input transformer or the differential input stage of an op amp.

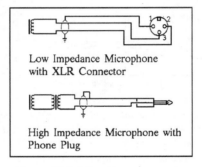

Low Impedance Microphone
with XLR Connector

High Impedance Microphone with
Phone Plug

Figure 11-3. *Microphone connections
will depend on whether the cable is a
balanced or an unbalanced line.*

Unbalanced lines (again, usually used with high-impedance mics) are shorter and have one side of the microphone output connected to the center conductor and the other side connected to the shield. While not as efficient at rejecting noise as balanced lines, they are much shorter (less than 20 feet), so they don't pick up as much noise anyway.

Standard specifications for low- and high-impedance microphones are about 250Ω for low-impedance and about 50kΩ for high-impedance. I have seen variations from these standards, though. There are lots of different kinds of microphones: dynamic, ribbon, condenser, etc. One of the most common types, the dynamic microphone is basically an inductive device. Sound waves acting on a diaphragm move a small coil in a magnetic field creating an electrical signal.

Most dynamic microphone coils are low-impedance. To convert them to high impedance, an impedance matching transformer is installed inside the microphone body. You can also get impedance matching transformers in separate packages with connectors on each end to allow easy conversion of existing microphones. Actually voltage output of a microphone varies from one type to the next and the sound pressure level it is picking up but is generally in the order of 10μV to 1mV.

Input Circuits

Microphone input circuits are set up to match impedances with the type of microphone connected. Older microphone inputs, (especially tube amps, but also some semiconductor circuits), used balancing and matching input transformers on balanced microphone inputs. Connecting the signal to the input of a transformer (which is isolated from ground) allowed the signal to be connected through to the amplifier while rejecting the noise present on both lines (because they were equal and opposite going into the transformer so they canceled).

Microphone preamps using operational amplifiers make use of the common-mode rejection characteristics of the op amp to cancel out the noise on both sides of the line. *Figure 11-4* shows a difference amp configuration that accomplishes this. If R1 equals R3 and R2 equals R4 the gain is simplified to the ratio of R2 divided by R1. As long as the resistor values are exact, and the op amp specifications are good, practically all of the common-mode noise signal should be rejected.

The input of the Rotel amp is a bit unusual. First, the low-impedance inputs accept balanced or unbalanced lines. You don't often see this on public address amplifiers. Analyzing the circuit suggests they have made some design compromises to accomplish this. The result is a more versatile input amp which probably doesn't reject common-mode noise as efficiently as it should. See *Figure 11-5*.

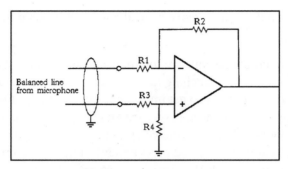

Figure 11-4. Microphone preamps using operational amplifiers make use of the common-mode rejection characteristics of the op amp to cancel out the noise on both sides of the line. This is the difference amp configuration that accomplishes this.

They've used a 4558 dual op amp and capacitively coupled the inputs, outputs and even the feedback loop. The two 680Ω resistors from each input terminal to ground create the input load impedance reflected back to the microphone. (The sensitivity of the microphone varies depending on the terminating impedance.) As *Figure 11-5* shows, the input amp is a difference configuration when connected in balanced mode. The unusual aspect of this design is that the resistors are not matched as in the previous example. This means common-mode signals will not be completely rejected. The reason for using this design is unclear but may relate to unbalanced operation.

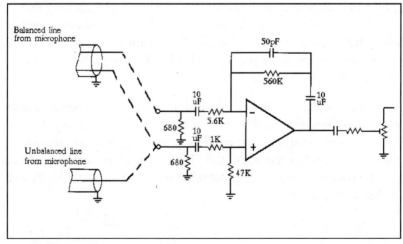

Figure 11-5. The input of the rotel amp is a bit unusual. The low-impedance inputs accept balanced or unbalanced lines.

When an unbalanced input is connected, the input to the op amp's inverting input is left open. Because the coupling capacitor is relatively large (10μF), it looks like a low resistance to all but the lowest frequencies. This has the effect of connecting that input to ground through the 680Ω resistor, creating a non-inverting configuration. The gain of this configuration works out to be about 90.

A switching jack changes the input amp into a "virtual" buffer amp (unity gain) when a high impedance microphone is plugged in. I say virtual because it connects 10μF into the feedback loop which, again, is a very low resistance for all but the lowest audio frequencies. The reason for this is the high-impedance microphone doesn't require as much gain as the low-impedance one.

The Public Address Amp

Before we talk about the output stage, let's see how the amp itself is laid out and discuss a few other features.

Figure 11-6 shows the block diagram of the Rotel amp. The tuner module has a terminal block that allows the connection of an antenna or cable. The output of the tuner is connected to the preamplifier but is also brought out to an RCA jack so you could connect it to another amp or other device. In addition to the microphone inputs to the preamp module, an auxiliary input and a phono input can be plugged into RCA jacks. The phono input accepts magnetic cartridge output levels and performs the standard RIAA equalization on the signal.

All of these signals are mixed together through resistors onto a bus before being applied to the next stage which incorporates the bass and treble controls. A muting circuit, consisting of a couple of NPN transistor switches, is connected to the point where the tuner, aux and phono signals come together. A contact closure across the muting input terminals effectively shorts these program signals to ground through one of these transistors. The microphones are unaffected by the muting function. By connecting a momentary switch on the paging mic to the muting input, paging can be accomplished.

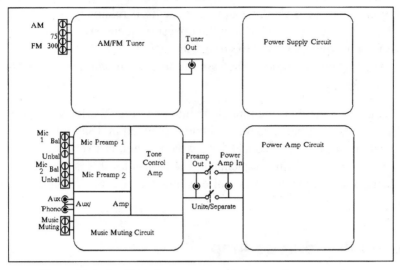

Figure 11-6. This block diagram shows the layout of a typical PA amp.

This amp also has the output of the preamp and the input to the power amp brought out to RCA jacks. The unite/separate switch allows you to disconnect the preamp from the power amp and use the jacks to insert an equalizer or other device. You could also mix in signals from other sources. This feature makes troubleshooting the amp easier because you can check for preamp output, and inject a signal into the power amp at this point.

The Power Amp

This amp, being a fairly new design, uses a monolithic power amp module in its output. The rated power output is 100WRMS, a respectable figure as public address amplifiers go. *Figure 11-7* shows a somewhat simplified diagram of the output stage, including the differential input stage made up of discrete transistors and the output transformer used to drive the speaker load. In past articles I've discussed quasi-complimentary transistor power amps and push-pull tube amps. This one really isn't all that much different in concept, just simpler because of the use of a chip.

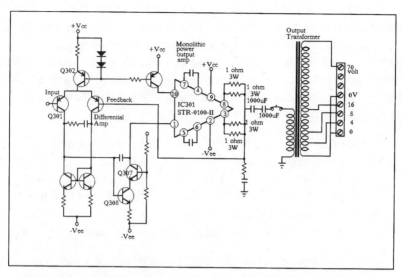

Figure 11-7. Simplified diagram of the output of a public address amplifier.

The input signal is fed into Q301 which is one side of the differential input stage. The other side of the amp comes from the output of the power chip and gives the negative feedback necessary to minimize noise and distortion, and maintain stability in the amplifier. Q302 is basically a constant current source to the emitters of the difference amplifier (diff amp). Its emitter current is set by the voltage across the two diodes connected to its base. The output of the diff amp goes through Q308 and Q307 before driving the chip.

Notice the two 1000µF capacitors in series with the output load (the transformer). These are good suspects if you're having trouble with an amp like this. Also keep an eye on the output resistors. They are 1Ω, 3W resistors and could go open in the event of a catastrophic failure. This amp is pretty forgiving (partly because of the transformer in the output), but nothing is indestructible.

The 70V Output

The final aspect of this system that I want to look at is the 70V output. This is probably the most characteristic—and least understood—aspect of commercial sound systems.

Seventy-volt lines are used in sound systems for the same reason high-voltage lines are used in power distribution systems. When you have to put in long speaker wire runs, losses in the wire become a problem. You want to deliver the same power to a speaker some distance away but you don't want high levels of current in your speaker lines. By increasing the voltage, the same amount of power can be delivered using less current. This means lighter gauge wire can be used.

But there is another advantage to 70V outputs on public address amplifiers. With a typical consumer or sound reinforcement amplifier the speaker load is usually pretty standard. The amp is rated to deliver some specified power into a specified load such as 16Ω, 8Ω, or 4Ω. If you make the load too small, the amplifier will exceed its maximum power output. The load is determined by the type of speakers and the arrangement in which they are connected.

These arrangements are standard (because you don't usually have very many speakers), but they're not very flexible. It doesn't take many 8Ω speakers connected in parallel to bring the load impedance below the minimum value. Series-parallel networks can be used but don't lend themselves very well to the physical layout of typical public address systems.

With public address systems—especially distributed, or ceiling speaker systems—you may want to connect many speakers to the system in a variety of locations. Usually you don't need more than a few watts out of each speaker because they are relatively close to the listener but twenty or thirty speakers wouldn't be uncommon in a medium-sized grocery store. The maximum output power of the amplifier is still the limiting factor in what you connect to the output but you need a way of doing it that doesn't cause problems for the amp.

The answer is to use standard 8Ω speakers with small impedance matching transformers which allow them to be connected to the 70 volt, high imped-ance speaker line. All speaker transformers in the system are connected in parallel across the 70V line making it simple to add speakers to the system. To control the output level of the speakers, the transformers have various taps to which to connect the speaker. The taps are labeled in terms of watts. In its simplest form, the procedure is to total up the watts of all the speakers connected. The total should not exceed the power rating of the amplifier.

Some Drawbacks

There is more to the determination of actual output and input impedances than I have mentioned but I've tried to concentrate on the practical side of it. In reality, 70V systems have some drawbacks. Often the quality of the transformers used degrades the frequency response out of the speakers. Inexpensive transformers also add insertion losses because they dissipate power themselves.

Older tube amps, which have inherently high output impedances, could often drive 70V lines directly. Semiconductor outputs are low-impedance, so a step-up transformer is generally used to create the 70V output. The transformer on the Rotel amp we've been looking at has multiple taps on its output. The lower taps match 4Ω, 8Ω and 16Ω speaker systems while the highest tap on the transformer drives the 70V line. Keep in mind, though, that only one output can be used at a time; either 70V or one of the others.

The other important thing to remember is that, when using the 70V line, all speakers must have impedance matching transformers. If you connect a standard 8Ω speaker to the 70V line the least you can expect is loud, distorted sound. The worst case scenario involves smoke billowing out of your amplifier. Rotel, in its wisdom, installed a circuit breaker between the power output and the transformer.

Well, there is so much more about public address that could be said. I couldn't hope to do it justice here but maybe this will help you decide whether you're interested in getting involved. Really, it's a field unto itself, related to other areas of electronics but with its own foibles and interesting technologies.

Chapter 12

Sound Advice

By John S. Hanson

The town library board called to invite me to bid on renovating the sound system in the library theater, which had been donated to the community by Baxter Pharmaceuticals. The library shows noon time movies to the senior citizens. There had been complaints about the sound. The budget for the renovation included funds for the purchase of both a video projector and a VCR.

Research For the Bid

To prepare for the bid I performed a sound level test using an audio test cassette and a Radio Shack 33-2050 sound level meter. The theater's projection booth had two Bell and Howell 16mm film projectors, a rack for the audio amplification equipment, and controls for the stage and auditorium lighting.

I inserted the audio cassette in the tape deck and turned it on, then moved from one seat to another taking sound level measurements in the 200 seat theater. I noted dead spots in the front center rows going back four aisles, and the music on the test tape sounded muffled (*Figure 12-1*).

Two column speaker systems were mounted up front in a vertical plane; one at each side of the stage some distance from the movie screen. The result was that in addition to the dead spots and muffled sound, the film sound appeared to be out of sync with the picture.

At this point I didn't believe that the problem was in the amplifier. Inspection revealed that it was a top brand commercial grade 100W model with separate mixer and equalizer.

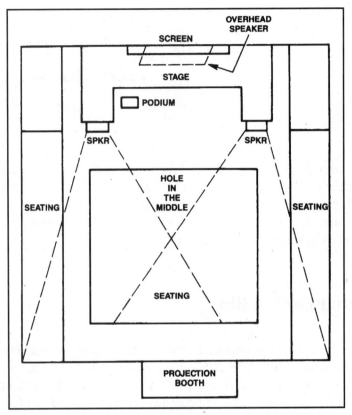

Figure 12-1. *The arrangement of speakers in Baxter Hall in the Morton Grove, IL library, connected out of phase, caused several sound system problems: muddy sound, a hole in the middle of the sound stage, and lack of synchronization between the film and the sound.*

My initial assumption was that the problem was caused by poor speaker placement, perhaps an impedance mismatch, or possibly an out-of-phase connection. Back at the shop I began writing my proposal.

Elements of the New System

At a trade show I had been impressed with the new LCD video projectors. I specified the leading maker's unit with a zoom lens and keystone correction that could be remotely operated. A wireless mic was a must, with the FM receiver in the projection booth. I preferred the hand-held model using a

174MHz carrier picked up by the six-inch antenna rod in the receiver.

Both the video projector and the VCR would be remotely operated from the front stage podium. As this distance is more than 30 feet, I included an IR repeater system. In this case, the IR remote data is RF transmitted on a 417MHz carrier, which is recovered and returned as IR data for commands at the receiver. Provided that the receiver is approximately ten feet away, the data will be in a 60-degree light pattern.

As I prepared the bid, my concern for the hole in the middle sound dispersion problem prompted me to specify an additional speaker system. Altec Lansing Voice of the Theatre speakers are installed directly behind the movie screen, which is porous, similar to a speaker grill cloth. I compromised with an over-the-screen installation with the speaker tilted down at 45 degrees.

Before I totalled my bid, I factored in my consulting time and technical labor time. My quote would assume I could use the library maintenance people and equipment.

I Got the Bid

The bid was presented and I awaited the pleasure of the library board. I knew of at least three competing bids. Weeks passed and I heard nothing. After some 60 days the call came saying that I had been awarded the contract. Before I started I made it clear that I wished to be paid along the way as the project progressed. This is an important clause you should include in any similar bids that you might make, to avoid a cash flow problem. The board agreed that I could invoice each month on the completed work.

Upgrading the Speaker System

First priority was the speaker system. The new speaker went in above the screen, and the column speakers remained in place. Using an old trick, I connected a 1.5V flashlight cell across each speaker system to confirm that all speakers were moving in the same direction when the same battery polarity was applied (*Figure 12-2*).

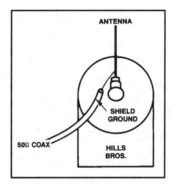

Figure 12-2. I used an old trick of connecting a 1.5V cell across the speaker terminal to determine phasing.

This test revealed that the column speakers were connected out of phase, explaining sound cancellation in some parts of the theater. I corrected this situation as I rewired the speakers. All three speaker systems were wired separately with 16-gauge lamp cord. The reason for separate leads was to have provisions for a future stereo connection. For now, the system would remain monophonic. I calculated the connection of the three speakers in parallel to present an impedance of 6Ω to the amplifier, so I used the 8Ω output tap on the power amplifier.

With the speaker system renovated, it was time to test the sound again. Using the same test tape, I sampled the theater with my sound meter and noted that the hole in the middle was gone, and all the seats had the same relative sound level. Making things even better was the fact that the music now had sparkle; the muddy sound was gone.

Installing the Video Equipment

Next came the installation of the video projector and the VCR. It was my understanding that cable operators generally provide free taps to public buildings, so I decided that I would ask the local cable company if they would provide cable at no charge for the library theater. The cable company brought in a cable tap for the VCR and tuner.

As I looked at the video projector I saw a problem. The IR remote sensor was mounted up front, under the lens. The VCR faced the opposite direction. I wondered if a mirror would effectively reflect the IR beam into the sensor. An 8 inch x 10 inch mirror placed under the lens about six inches from the sensor did the trick. Needles to say, I was elated. Backing off, about 10 feet in the booth, I was able to operate all functions of both the projector and VCR remotely.

After turning on the projection TV and adjusting the keystone screen size and focus, I saw the brightest, sharpest video display on a movie screen that I had ever seen. The picture rivalled the brightness on the film projectors with no convergence problems and no arcing and slumping of CRTs. I was truly impressed.

Correcting Some Audio Problems

The audio posed an immediate problem. With the wireless mic on, a strange sound emanated from the speaker system. It sounded like a big windshield wiper blade swishing in the rain. As it only occurred when the video projector and wireless mic receiver were both on, I concluded that the problem was pickup of switching harmonics from the LCD panel pixel switching pulses. After some testing, I concluded that I needed a shielded antenna input and an external antenna for the mic receiver.

First I removed the rod and antenna and hard wired a length of 50Ω coax right to the PC board. A metal coffee can lid would make a perfect antenna mount and ground plane (*Figure 12-3*). Mounting the antenna rod to an insulator, and then to the can lid, I soldered the other end of the coax to the antenna rod and the shield to the can, disappearing into the attic to work my way back to the area over the stage.

Fiber glass insulation penetrated my pores. I was miserable. Some 80 feet from the projection booth I left my Hills Brothers antenna and followed the coax back. In the booth, the mic worked perfectly with no sign of any interference. I later learned that this is a common problem in Las Vegas. Techs even use miniature Yagi type antennas to eliminate signal drop outs caused by interference.

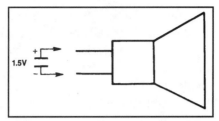

Figure 12-3. In order to eliminate audio interference caused by switching in the LCD projection TV, I used a coffee can and lid to provide grounding and shielding for the wireless microphone system.

This Problem Would Affect the Remote Repeater as Well

Now I was starting to think that I would have the same problem with the remote repeater. The IR transmitter unit was placed about ten feet from the podium allowing a 30-degree acceptance pattern on both sides of center. Watch out for daylight and interference from quartz spot lights. Both can cause IR crosstalk and can blank the remote's transmission.

In the projection booth I mounted IR repeater translater receiver in the back corner near the ceiling. The beam operated both units thanks to the mirror. When the interference problems that I had anticipated did not surface I was delighted.

Time For the Demonstration

The library board came in and took random seats in the theater as I used the wireless mic to demonstrate complete remote operation from the podium. I powered up the video projector, turned on the VCR, started, paused and stopped the video tape. For an encore, I mode switched the VCR to access a cable TV program. the board was delighted. My check for the balance came in the mail without delay.

Conclusion

With consumer electronics sales service becoming less profitable, it's time to explore commercial electronics. My library story is an example of a profitable project that was fun.

My next project will be a sports bar. Television started in the bars, saloons and taverns and it's coming back. This time it's with big bucks installation complete with a satellite dish and multiple monitors. The bar has gone high tech. Sell now.

Learn how to bid successfully. Don't sell the worth of your time short. Price becomes less of a factor once you established a can-do reputation.

Chapter 13

Understanding Compact Disc Troubleshooting Concepts—Part 1

By Marcel R. Rialland

Compact disc technology was introduced in 1982, with the idea of greatly improving audio recordings. The compact disc has now become a widely accepted medium for audio recordings. Because of its large storage capacity, the compact disc is now used for applications other than audio; including CD-ROM (compact disc read-only memory), photo-CD, and CDI (compact disc interactive).

Laser Vision is also becoming more popular and new developments in optical recording devices, such as Sony's Mini Disc, have recently been introduced. All this new technology means expanded opportunities and challenges for the service technician.

The Troubleshooting Challenge

Troubleshooting compact disc players (or other laser read and write systems) can be quite a challenge. A thorough knowledge of the theory of operation can help a technician become more effective in diagnosing compact disc player faults. Also, there are some tools, test jigs, test discs, and test equipment (oscilloscope and DVM) that are needed to simplify service procedures.

The required equipment is usually listed in the service manual; which is also needed to service a compact disc player. The service manual contains service procedures, such as how to enter the test modes that assist in diagnosing the player. In many players, the test modes are used to check the condition of the laser pen (optical pick-up unit) as well as the status of the servo circuits.

Basic CD Operation

The simplified block diagram, shown in *Figure 13-1*, illustrates the process of reading and decoding a compact disc. This process must be followed in all CD systems. First there must be a means of retrieving the data from the disc. This is accomplished by the optical pick-up unit (OPU), which is part of the CD mechanism. The CD mechanism provides the means for the OPU to track the disc.

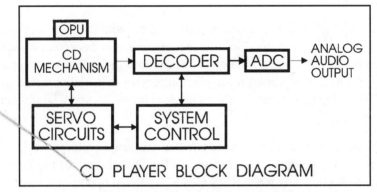

Figure 13-1. CD player block diagram.

Normally the spindle or turntable, which is used to spin the disc, is also part of the CD Mechanism. The position of the OPU, focus, tracking, and the data retrieval rate (spindle motor speed) are controlled by the servo circuits. The system control block controls the status of the servo system (including the start-up sequence), and interprets and initiates commands received via the remote control or local keypad.

The data from the disc must also be decoded. At the least, decoding includes: EFM demodulation, deinterleaving, error correction, interpolating unrecoverable data, digital to analog conversion, and filtering. Most of these processes are performed by the LSI ICs and there is little that a technician can do about what goes on internally. However, there are signals (clock and data) as well as control lines that one should be familiar with in order to become proficient in troubleshooting the CD player.

The data includes more than just digital audio samples. The data also includes parity bits (data for detecting errors), and control and display symbols. The decoder deinterleaves the data and uses the parity bits to recover unreliable data.

The control and display information is retrieved and used by the system control circuit for controlling the CD player system, and for displaying data related to the disc that is playing, such as elapsed time and the current track. The right and left audio 16-bit samples are converted to analog audio by the analog to digital converter (ADC).

Two Types of Servo Systems

There are basically two types of servo systems, the single beam system and the three beam system. Most CD players contain three servo loops: the spindle (turntable) servo, focus servo, and tracking servo (radial tracking).

In addition, the laser beam intensity is controlled by a feedback control circuit. The servo loops must be operating before the data from the disc can even be retrieved (read) and decoded. The three beam system is examined in this article.

The Three Beam Servo System

The bottom view of a CD mechanism for a three-beam servo system is shown in *Figure 13-2*. The CD mechanism can be divided into three main sections corresponding to their respective servo circuits: focus, turntable and tracking.

Figure 13-2. Three-beam servo CD mechanism (bottom view).

The focus servo circuit controls the focus movement of the objective lens (the objective lens is not shown).

The turntable servo (TT servo) controls the speed of the turntable motor.

As the compact disc spins, the tracking servo causes the OPU to follow the track. The tracking servo controls the position of the OPU in two ways: one for small (high frequency) radial corrections caused by the eccentricity of the disc and one for greater (low frequency) tracking corrections as the OPU tracks across the disc.

The first servo controls the movement of the OPU via radial tracking coils which control only the radial movement of the OPU. This loop allows fast response to the tracking servo. The second moves the entire OPU assembly on a sled via a stepper (sled) motor and gears.

The Optical Pick-up

The optical pick-up unit is a complex electromechanical device designed to optically read and track the compact disc's spiral tracks. As illustrated in *Figure 13-3*, the three-beam optical pick-up unit is generally composed of lenses, mirrors, a laser diode and photodiodes.

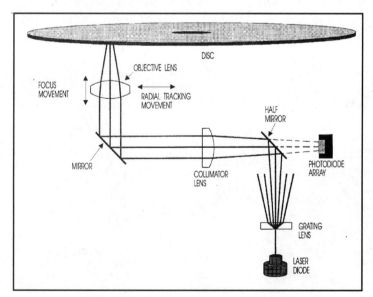

Figure 13-3. Three-beam optical pick-up unit.

A single beam, emitted from a laser diode, is split into several beams of which only three are used; the three in the center. The center or main beam is the most intense beam used for focus and reading the pits on the disc. The secondary or radial beams are used strictly for tracking.

The light bundle (three beams) is reflected by the half mirror toward the collimator lens (all three beams are placed in parallel by the collimator lens). The beams are then directed by another mirror up through the objective lens.

When the player is placed in a service mode to initiate a start-up, the lens can be seen moving up and down several times. The OPU should also move inward to locate the lead-in track.

If a disc is detected, the light bundle (modulated by the pits in the disc) is reflected back through the objective lens, mirror, and the collimator lens. Some of the reflected, modulated light bundle passes straight through the half mirror to strike the photodiode array.

Alignment is Critical

The alignment (grating angle) of the three beams is very critical in the three-beam optical pick-up system as shown in *Figure 13-4*. An improper grating angle results in poor tracking or, in severe cases, tracking may not be possible at all.

In many cases, the grating angle is factory set and cannot be adjusted. Where there is an adjustment, the manufacturer's grating angle adjustment procedure as outlined in the service manual must be followed. Normally the adjustment is only made when replacing the optical pick-up unit.

The three beam servo block diagram is shown in *Figure 13-5*. The photodiodes provide focus and tracking error signals to the focus and tracking servo circuits. The focus error signal is developed by the focus servo circuit and is fed to the focus drive circuit. The focus drive provides the drive signal to the focus coil on the compact disc mechanism (CDM) to keep the laser beam focused on the disc.

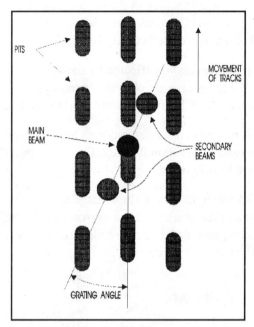

Figure 13-4. *Grating angle.*

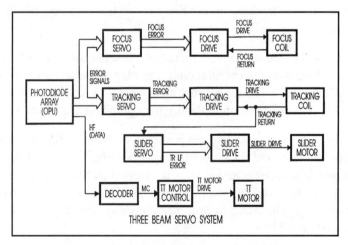

Figure 13-5. *Three-beam servo system block diagram.*

The Tracking Servo Circuit

The tracking servo circuit similarly receives tracking error signals from the photodiodes and develops the tracking (TR) error signal. The tracking error signal is fed to the tracking drive circuit, which applies drive to the tracking coil on the pick-up mechanism.

The coil provides the tracking (TR) return to the tracking drive and to the slider servo circuits. The slider servo provides low frequency (LF) tracking corrections. The TR LF (tracking low frequency) error signal is sent to the slider drive circuit to drive the slider motor.

The rotational speed of the turntable (TT) motor is controlled by detecting the data (detected from the HF) coming into the decoder block. This is done in the process of bit clock regeneration and decoding the incoming data. The decoder develops the motor control (MC) signal to regulate the speed of the motor and thus keep the demodulated data coming at the correct speed (4.3218 Mb per second).

Starting the Disc

Every CD player must go through a start-up procedure in order to start playing a CD. It is important to know the start-up sequence when trouble-shooting a condition where the disc will not play. The start-up sequence is initiated by the microprocessor when it receives a command to detect the presence of a CD or to start playing the disc.

The start-up initiates as follows (note: some of the steps occur simultaneously):

1. Turn the laser diode on.
2. Focus the laser beam (the objective lens will move up and down until focus is achieved).
3. Start the disc turntable motor (TT servo locks when data is detected).
4. Move OPU to the lead-in track and read the table of contents.

Although there are variations in the start-up sequence between makes and models, all CD systems must find focus and find the control and display data (especially the table of contents in the lead-in track) to play a disc.

Once the CD starts, all the servos remain locked, until a jump track command (skip track or fast forward or reverse) is received. In this case, the servo microprocessor works with the tracking servo to allow skipping of tracks until the desired selection has been located.

A laser power meter can be used to check the status of the laser beam. The laser power meter is also used to adjust the laser power level following the service manual's instructions for the unit under test. This check or adjustment is usually performed while the player is in the service mode.

In some players, the intensity of the laser diode is measured by checking a reference voltage in the laser diode control circuit. This voltage is proportional to the intensity of the laser diode.

Part two will take a closer look at some troubleshooting techniques for the three-beam servo circuit and introduce the single-beam servo system.

Chapter 14

Understanding Compact Disc Troubleshooting Concepts—Part 2

By Marcel R. Rialland

As considered in Part 1 of this series, it is important to understand how a product works before attempting any troubleshooting. In this part, the single-beam servo system will be examined and common troubleshooting techniques will be applied to both the single-beam and three-beam systems.

In many respects the single-beam and three-beam systems are quite similar. But there are some important differences which should be considered. The major difference is in each mechanism's tracking system. The difference is discernible in looking at the CD mechanism itself.

Actually, the single-beam system's CD Mechanism (*Figure 14-1*) is simpler. Rather than using a sled assembly to track the disc, the single-beam system uses a radial swing arm, which pivots on a single axis to radially follow the spiral tracks.

The Single-beam Servo System

In order to understand the single-beam servo system we need to examine the optical pick-up unit more closely (*Figure 14-2*). Just as with the three-beam system, a single laser diode is used to emit a laser beam. The single laser beam is directed to the disc by the half mirror and the collimator lens and is focused via the objective lens.

This single beam must be controlled to read, focus, and track the disc's spiral tracks. The spinning disc modulates the single beam as it is reflected back into the objective lens. The reflected, modulated beam is directed back to the photodiodes. In order to detect focus and radial errors the beam is split to strike two areas of the photodiode array as illustrated in the photodiode array physical layout of *Figure 14-3*.

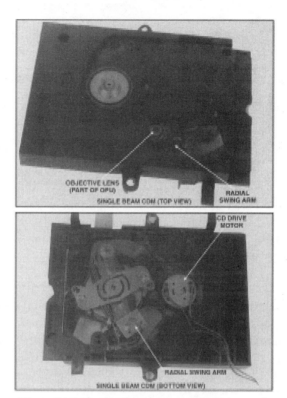

Figure 14-1. *View of the single-beam CDM (top and bottom view).*

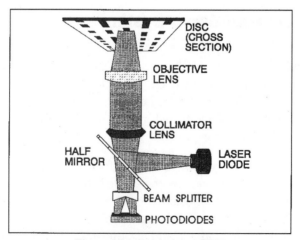

Figure 14-2. *Single-beam OPU.*

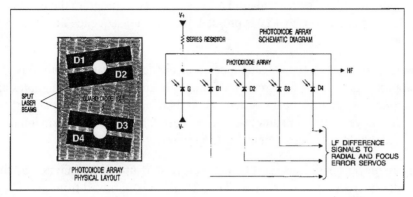

Figure 14-3. Photodiode array.

The photodiode array contains up to five photodiodes as illustrated. The output from each photodiode is totaled as the HF signal (IHF=ID1+ID2+ID3+ID4+IG). Tracking and focus errors are detected by calculating the differences in the photodiode low frequency (LF) currents.

Radial (tracking) errors are detected by calculating the difference between the total current of D1+D2 with the total current of D3+D4, which can be expressed as IRE=(ID1+ID2)-(ID3+ID4).

Focus errors are detected by calculating the difference between the total current of D1+D4 with the total current of D2+D3, or IFE=(ID1+ID4)-(ID2+ID3).

These signals are processed by the servo circuits to provide control of the CD mechanism.

Start-up

In most players, each time a CD is placed in the player, a start-up sequence is initiated to read the table of contents from the CD. In this way, the player detects not only the presence of a CD, but it also determines the type of disc present and decodes information pertaining to that disc. Some of that information is decoded and displayed on the CD player's display.

Some of the data is also used by the microprocessor to control access to the tracks on the disc. The information allows the user to program the tracks to be played. When the play key is pressed, the start-up sequence must again be initiated.

Figure 14-4 shows a typical single beam laser/focus start-up circuit. Although this is the start-up circuit for a single-beam system, the principles can be applied to the three-beam system. Also there are variations in the start-up circuits between different single-beam models. Check the service manual for the model being serviced when checking the start-up circuit.

Certain player conditions must be met before start-up can be initiated by the decoder microcomputer. When a CD mechanism is removed from the cabinet to allow troubleshooting, these conditions are at times difficult to accommodate. For example, some players have tray switches to signal the microcomputer that the tray is closed. This switch must be closed before you attempt start-up troubleshooting.

In the stop condition, SI/RD (IC6501, pin 6) is low. Start-up is initiated when the photodiode signal processor (IC6501) receives a start initiate signal (high) from the decoder microcomputer (IC6530) via pin 6. The start capacitor (C2513) begins to charge as indicated by the start-up signals. At that time pin 17 (low) supplies about 3V to the laser driver circuit to turn the laser diode on.

The focus search is initiated by swinging the FE voltage between +1.2V and –1.2V. This causes the objective lens to move up and down to attempt focus servo lock. The focus search pattern occurs twice in this model. If focus is not found after the second focus attempt, the system assumes there is no disc present.

Start-up Problems

In troubleshooting start-up problems, the start-up signals can be checked during the start-up initiation. Using an oscilloscope, with the input set to dc, allows the start-up initiation voltages to be observed.

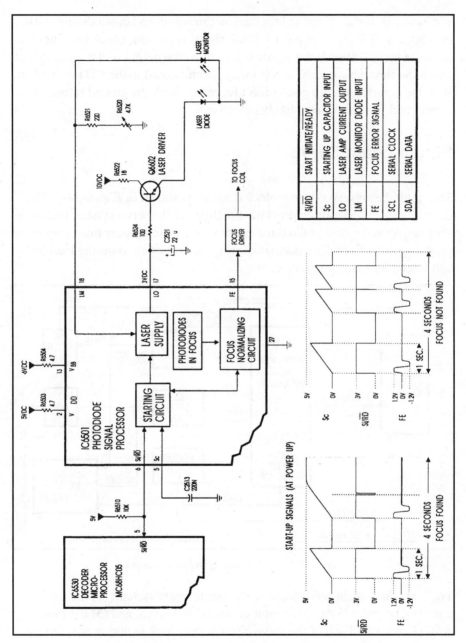

Figure 14-4. *Focus/laser start-up circuit.*

For example, if the objective lens does not move up and down during start-up, check the FE signal at pin 15. If the signal is present, check the output from the focus driver. If the signal is present at the output of the focus driver, check the flex cable connection from the circuit board to the CD mechanism. If all the connections are good (don't forget to check the ground return), the optical pick-up unit is most likely defective.

Single-beam Servo

The single-beam servo system block diagram is shown in *Figure 14-5*. The decoder microprocessor controls the functions of the servo system, including start-up. As is the case in the three-beam system, the decoder microprocessor must also control tracking during some operations, such as during track loss and search forward and reverse.

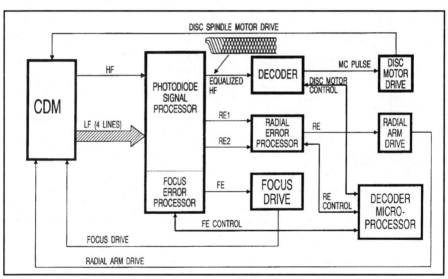

Figure 14-5. Single-beam servo block diagram.

There are three main servo loops in the single beam system: (1) the focus error (FE) servo, (2) the disc motor control (MC) servo, and (3) the radial error (RE) servo. Each of the servo circuits must lock in during the start-up procedure in order for a disc to play.

First the laser diode must come on. Next, thelaser beam must be focused onto the CD, which also indicates the presence of a disc. Then the disc motor spins the disc and the radial arm pivots to locate the lead-in track in order to read the CD's table of contents (TOC). The TOC contains information about the disc: the type of disc (CD-DA, CD-ROM, or CDI), total number of tracks, total time, and the CD's catalog number.

After the TOC is read, most CD audio players display the user information, such as the total tracks and total time, on the front display. Therefore, the display is a good place to start looking for clues when troubleshooting.

For example, if disc information is never displayed, there may be a problem in the start-up. If there is a disc error during start-up, monitor the equalized HF signal (eye pattern) output from the photodiode signal processor during the start-up procedure. The clarity of the eye pattern can give a clue as to where the problem may be located.

Of course, if the HF signal is never present, the problem may be in the start-up initiation. For example, the laser may not even be coming on. The service mode can reveal if this is the case. Moreover, a laser power meter can show if the laser diode is indeed coming on.

Focus Error Servo

Although the objective lens and the focus drive circuit of each system (single-beam and three-beam) are similar, the focus servo of the single beam is quite different from the focus servo of the three-beam system. The four low-frequency signals from the CDM are processed by the photodiode signal processor to develop the focus error signal.

Internally, the LF signals are applied to adders and comparators to find the focus error (FE is actually a focus error correction signal to keep the laser focused). Generally, there is a focus drive circuit that applies the focus correction drive signal to the focus coils on the optical pickup unit.

Most CD players require some adjustments (such as focus offset and laser current) in the focus servo loop for optimum performance. If the focus offset adjustment is too far off, the disc may have trouble tracking and may even stop playing due to error occurrences during the decoding process. Observing the eye pattern of the HF signal can help in determining if there is a problem in the optical pick-up system. The eye pattern should be fairly clear as shown in *Figure 14-6*.

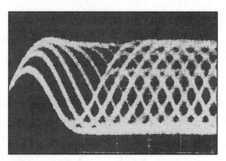

Figure 14-6. Equalized HF (eye pattern).

Disc Motor Control Loop

The method of controlling the disc rotational speed is similar to the three-beam system's method. The disc motor control servo loop controls the rotational speed of the disc motor. As shown by the block diagram, the HF is amplified and equalized by the photodiode signal processor. This equalized HF signal, which is observable as the eye pattern at the output of the photo-diode signal processor, is applied to the decoder block.

The decoder is usually an LSI IC which performs several functions for decoding the incoming data (e.g., data slicer, demodulator, EFM decoder, interleaver corrector and descrambler, and interpolator). In the decoding process, the bit clock of 4.3218Mb/sec must be regenerated.

At the same time the incoming data flow must be regulated so that the data does not come in too fast or too slow. The regulation is provided by means of the pulse width modulated motor control (MC) pulse, which is used to control the rotational speed of the disc spindle motor via the disc motor drive block (integrator circuit).

Starting and stopping the disc motor is controlled by the decoder micropro-cessor via the SSM line. If the decoder does not receive the control information from the decoder microprocessor, the disc spindle motor may not start; or, the opposite may occur. The spindle may start but not stop when a stop is initiated.

Radial Servo Loop

Just as in the three-beam servo system, a radial servo loop is necessary to provide proper tracking of the disc in the single-beam system. As shown by the block diagram (*Figure 14-5*), the radial error correction signal is primarily developed in the radial error processor.

The radial error (RE) correction signal is developed from RE1 and RE2. RE1 and RE2 are derived from the LF signals, which are processed in the photodiode signal processor. RE1 is the sum of the currents from D1 and D2, and RE2 is the sum of the currents from D3 and D4.

In addition to these signals, there is also a wobble signal of about 650Hz which is injected into the radial servo loop of the single beam system. This wobble signal is generally produced by an oscillator in the radial error processor. If this signal is not present the disc will not track properly (poor playability).

The wobble signal is introduced into the service loop to compensate for laser spot asymmetry errors (D-Factor) and to compensate for the tracking angle variations as the radial arm moves from the inner tracks to the outer tracks of the disc (K-factor).

Troubleshooting Summation

As with any consumer electronics product, it's a matter of isolating the symptom to a particular circuit. Then, with a few more checks, the problem can be traced to a few components. These components can then be checked or substituted to verify which component is indeed defective.

Knowing the operating requirements of the CD player is essential to diagnose a symptom properly. It's a matter of determining the path to follow. For example, if power is applied to the CD player and the display doesn't illuminate and no other activity is observed, the power supply is the most likely place to start.

If, however, the display does illuminate and other activity is observed, the power supply can most likely be ruled out. But even in this case, a problem may still be traced back to the power supply.

Let's apply this troubleshooting technique to both the three-beam and single-beam servo systems. The symptom is: "the disc will not play."

First, look for the start-up condition (you may want to use the service mode to verify the condition of the servo system):

1. Does the laser diode come on? If not, check the laser current source.
2. Is the focus search initiated (the objective lens moves up and down)? If not, check the start-up initiation drive signals with an oscilloscope.
3. Is focus achieved?
4. Does the OPU move to search for the lead-in track?
5. Is the lead-in track found (table of contents track)?

If all these processes are working while in the service mode, the CD player's servos should all be locked. The eye pattern (HF signal) should be available to the decoder circuits. Verify this. There may be a problem in the decoder section.

If, for example, the decoder's subcode processor was not functional, the display may show an error and the disc would go into the stop mode, which may appear to be a no start-up condition. But the problem may be due to a decoder fault.

Diagnosing Poor Playability Symptom

First, let's define poor playability. Poor playability can be poor tracking, poor tracking on only some discs, difficulty in start-up, easy loss of track due to jarring.

What can cause poor playability? Don't overlook the obvious: a dirty objective lens reduces the optical system's efficiency and can even prevent start-up. A little alcohol on the end of a cotton swab can be used to clean the objective lens. However, do not apply too much pressure in cleaning the lens.

If the lens is clean, make sure there is nothing obstructing the mechanical tracking mechanism (both in the three-beam sled system and the single beam radial swing arm). If it is a single beam system, make sure the wobble signal is present.

Do not overlook the possibility that there may be more than one symptom. For example, I came across a CD player (single beam) which displayed two symptoms: the radial arm swung all the way to the outer perimeter and the objective lens was pulled to its lower extremity. I traced the problem to an open resistor from the power supply to a quad op-amp, which was common to both the focus drive circuit and the radial drive circuit.

Chapter 15

What Do You Know About Electronics? Mostly Radio

By Sam Wilson

I get many publications and advertisements in the mail every month. I try to seek out subjects that I think will interest ES&T readers. In this issue I have included information I have received in the past two months.

No Broadcast Radio Operators Need Apply

According to an article in Radio World, the FCC has waived its rules in order to allow unattended operation of radio broadcast stations. Fully-automated stations are now possible. This move was requested by the NAB (National Association of Broadcasters).

The ruling eliminates the need for a Restricted Radio Telephone Operator Permit. Not much has been eliminated here. The "test" didn't really test very much. The hardest part was remembering your name and address.

History Lesson

On a recent news broadcast I heard that over 50% of high school graduates do not know history. That is supposed to be news? I'd like to know what kind of questions they asked to test history knowledge. What did they consider to be the important dates and events on the test?

When I went to school they left out a lot of the really good stuff in history class; like who was R.A. Fessenden, and what great thing did he do in 1902?

Of course, every reader of WDYKAE? could answer that one. Fessenden was the first to send music by radio. The music was received at a distance of 48 miles.

RDS and RBDS

RDS (Radio Data System) has been in operation in Europe for a long time. The U.S system, RBDS (Radio Broadcast Data System), is about to be launched. Both systems allow FM stations to transmit data to anyone with a receiver equipped to receive it. A few examples of how it can be used are given here:

• Promotional information (such as the name of the recording company and rating of the songs and music being played).
• The name of the artist(s).
• Title of song(s) and music being played on the station.

Stations can earn additional income by using the extra RBDS system for paging, and leasing the system to companies with special applications. One example is the use of the RBDS system to broadcast stolen or lost credit card information.

Figure 15-1 shows the spectrum of the FM station with RBDS and the newly-allotted dual SCA (subcarrier authorization) services. The allotment for RBDS is a low-amplitude, narrow-band signal.

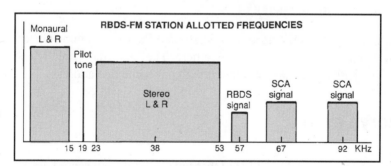

Figure 15-1. *This is the spectrum of the FM station with RBDS and the newly-allotted dual SCA services. The allotment for RBDS is a low-amplitude, narrow-band signal.*

Radio listeners who want to adapt their current radios to RBDS will need an add-on unit that displays the added information in an alpha-numeric readout. The station can use 16 different data groups. Each one has four blocks of information that is 26 bits long.

There are 16 possible data groups available, but only 13 have been assigned. Your job, should you decide to accept it, is to come up with an exciting new application for the remaining groups.

There's another million dollar idea.

One application of RBDS allows the consumers to select their favorite entertainment format. The RBDS equipped radio will page through stations and select one with the desired format (talk show, news, rock n' roll, whatever). There are over 30 program types to choose from.

Maybe, just maybe, you will find something you like on the radio.

Who Will Be Able to Service These?

A company called ROLLS Corporation of Salt Lake City, UT, has announced a new audio system. Their model RP220 provides clear sonic quality and performance for the working musician, studio engineer, or anyone who needs a preamp with a "warm, smooth, analog sound."

The RP220 uses 12AX7A tubes in a unique configuration to give smooth, controllable gain with true transformer balanced inputs. The RP220 has several inputs and outputs, and MIC/ line switches on the inputs and outputs to adapt to any performing or studio situation.

Why Tubes?

The primary purpose of a preamp is to provide low-noise gain to microphones and instruments without coloring or distorting the sound in any way. But any audio electronic device seems to have a characteristic "sound." Sometimes that sound can be pleasing, sometimes not. Tube preamps have gained in popularity recently because some audiophiles do not like the so-called "digital sound" of digital recording equipment.

Many engineers prefer the "sound" of analog recording equipment but like the editing and predictability of digital. Because tubes have some natural compression before they clip, and when they do clip they have a more even harmonic structure, they have a smoother sound than solid-state preamps.

Attempts to duplicate these characteristics with solid-state circuits have had limited success.

Specifications for the RP220:

•Frequency Response: 20Hz to 20KHz, +1dB

•Input Impedance: 600W balanced low Z, 10kW line, 1MW instrument.

•THD: 0.05% typical

•Max Gain: 40dB Instrument or line, 60dB low Z

•Indicators: 5-segment output level, +48V LEDs, Power Status LED

•Dimensions: 3.5" x 6" x 19" (89mm x 162mm x 482mm)

•Power: 120Vac (230Vac) 15VA.

Speaking of FM

This is a good place to review some basic FM terms.

The frequency deviation (DF) of an FM signal is the amount of frequency change above and below the carrier (center) frequency. Deviation is measured in Hertz.

The modulation index (ß) is the ratio of the frequency deviation to the audio frequency that produces that deviation.

The frequency deviation and modulating frequency are related to the modulation index by the following equation:

frequency deviation modulation index = modulating (audio) frequency

In symbols, that would be:

DF ß = DF/Fm

The sidebands of an FM signal can reach out a great distance from the center frequency. A practical measurement is to include only the sidebands that have an amplitude of 10% or more of the center frequency. The usual equation for the bandwidth is:

bandwidth (BW) = 2 x FM (1+ß)

Vocabulary Time

Vocabulary is an important part of your electronics knowledge. O.K., so you know:

• what a goniometer is used for,

• what an Adcock antenna looks like,

• what a vector diagram of the coriolus effect on a 600mph jet traveling west-to-east looks like.

Knowing those definitions might get you points on next month's Test Your Electronics Knowledge.

But what if someone asks you to define thermography? Can you do it? (My Dictionary of Technical Terms spells it thermiography, but a company that makes them spells it thermography).

Sure you can define it. You know that everything in the universe radiates heat at temperatures above absolute zero. You also remember that the heat is radiated in the form of infrared radiation. And you know that thermography is the technology of measuring temperature at a distance by measuring the amount of infrared radiation from a body and converting it to degrees temperature.

Plastic Cable

Not all fiber optic cable is made with glass. DuPont developed a plastic fiber optic cable called Crofon that has been in use since 1969. It can be bought in a 1mm (0.0394 inch) diameter cable encased in a 2.2mm optical cladding.

Joe Risse sent me the plans for using any fiber optic conductor as a burglar alarm (*Figure 15-2*). The idea is to thread the cable through your computer, printer, tower, etc., then connect it via the appropriate optical interface to the alarm. This alarm system can also be used to safeguard your stationary test equipment.

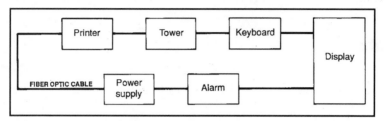

Figure 15-2. Here's a way to use fiber optic conductor as a burglar alarm: thread the cable through your computer, printer, tower, etc., then connect it via the appropriate optical interface to the alarm. This scheme can also be used to protect your stationary test equipment.

When the burglar cuts the fiber optic conductor it breaks the light beam and sets off the alarm. You can design this simple system yourself. Use their CLOE-1040. Get the specs from:

Crofon Marketing
774 Limekiln Road
New Cumberland, PA 17070

Chapter 16

Commercial Radio Licenses: A Step on the Road to Success

By Dale C. Shackelford

What do a cruise ship officer, an international airline pilot and an electronics servicing technician have in common? They all need a commercial radio license to make the most of skills they have attained to enhance their careers (and incomes). The type of commercial radio license one might need/desire will often vary with individual goals, but for those of us in the electronics servicing field, the General Radiotelephone Operator License (GROL) and the Global Maritime Distress and Safety System Maintainer's License (GMDSS/M) are just the tickets.

Why would I need a GROL or GMDSS/M license?

As of June 15, 1984, the Federal Communications Commission (FCC) requires anyone who repairs, maintains and/or calibrates any ship (including marine and aircraft), coastal station or portable marine band radio, to hold, at a minimum, a GROL. Additionally, anyone who maintains or repairs any AM, FM, TV or international broadcast stations (including short-wave), auxiliary broadcasts/services (low power TV, FM or TV broadcast translators, boosters, etc.), or any fixed radio telephone/radiotelegraph stations, must hold (at a minimum) a valid GROL. Technicians who desire to repair, maintain or calibrate any of the new satellite based marine emergency subsystems or equipment must hold a GMDSS/M, while those who desire to repair, maintain or internally calibrate ship radar systems must hold a GMDSS/M with a Radar Endorsement. Obviously, there is a lot of work for technicians who qualify for these licenses.

What Are the Qualifications For a GROL Or GMDSS/M?

To qualify for a GROL, you must: be a legal resident of the United States or otherwise be eligible for employment in the US, be able to receive and transmit spoken messages in the English language, and be able to pass written examinations covering basic radio law and maritime procedures (FCC Element 1) and electronic fundamentals and techniques (FCC Element 3).

For the GMDSS/M license, the applicant must pass an additional FCC element (9), consisting of 50 questions on general radio maintenance practices and procedures, of which, 38 (or 75%) must be answered correctly. Applicants taking tests on Elements 1 and 3 must also correctly answer 75% of the questions to pass the specific element, otherwise the test will have to be readministered.

Element 1 consists of 24 written questions (requiring 18 to be answered correctly) while Element 3 consists of 76 written questions (in 8 sub-elements), requiring that 57 be answered correctly before passing the test. Applicants must also pay all applicable regulatory fees and tees that COLEM may charge (see below for the definition of a COLEM).

Upon meeting all of the criteria (as set forth above), the passing of each individual Element will qualify the applicant to hold the permit (or license) covered by that particular Element. For example, if the applicant took the test for the GMDSS/M (Elements 1, 3 and 9) but only passed Element 1, the applicant could accept a Marine Radio Operator Permit (MROP), which requires passing only Element 1. The applicant could then retake Elements 3 and 9 at a later date to qualify for the GMDSS/M, without having to retake the Element I test. Alternatively, the applicant could accept a PPC (Proof-of-Passing Certificate) as described below.

What Exactly is the GMDSS?

The Global Maritime Distress and Safety System is an automated ship-to-shore distress alerting system using satellite and other advanced (terrestrial) communications systems which will eventually replace Morse code as a maritime distress communications medium. This system, coordinated world-wide by the International Maritime Organization (IMO) provides rapid transfer of a ship's distress call to the agency (Coast Guard, Civil Air Patrol, search and rescue, etc.) best suited to provide the necessary assistance in an emergency. The GMDSS allows each station to be assigned a unique call signing a system which has been allocated a select band of frequencies upon which to operate worldwide. To operate a GMDSS system, one must hold a GMDSS/0 (Operator) license, while one must hold a GMDSS/M to maintain, repair or calibrate these systems.

Who Administers Commercial Radio Operator License Tests?

In October, 1992, the Federal Communications Commission (FCC) transferred the responsibility of commercial radio license testing to the Private Radio Bureau, the same entity that handles Amateur Radio operator examinations.

The new commercial radio operator testing program is currently being directed by nine (private) primary organizations known as Commercial Operator Licensing Examination Managers (COLEMS). Each COLEM may have any number of testing facilities across the country,though each facility will be responsible to, or licensed under the authority of, the primary COLEM. You can find the facility nearest you, as well as testing dates and costs by contacting the COLEM of your choice.

Upon passing each element on the road to receiving a GROL or GMDSS/M, the COLEM test administrators will complete a Proof-of-Passing Certificate, noting the element passed, identity of the applicant and the date passed. This will allow the applicant an entire year to complete other elements without having to retake the passed element test(s) or apply for an "inferior" class of license/permit.

Studying For the Commercial Radio Operators License

Once every few years, the Federal Communications Commission releases into the public domain a set of question pools for the various elements of the commercial radio licenses (including, but not limited to MROP, GROL and

GMDSS/M).These pools contain every conceivable question (and answer) which could be asked on any of the FCC required/COLEM administered element tests, as all test questions are required by the FCC to be culled from the question pool for that specific element. While these question pools do have all of the (multiple choice) questions and their corresponding answers, there are no explanations as to why the answer may be right or wrong.

To fill the void left by the question pools released by the FCC, many COLEMs, in addition to administering commercial radio tests, sell study guides or hold classes (probably in your area) which will help applicants comprehend the information required to pass a specific element(s). These study guides are often available in book or computer software form, and are well worth the investment, regardless of how experienced one is in the field of electronic repair.

Although not required to do so, COLEMs can also provide invaluable assistance in filling out the proper forms (in the proper manner) for submission to the FCC. These forms contain various codes for things such as fees, which are not readily evident or self explanatory, and can be extremely intricate. If a form is filled out wrong (or incomplete), the FCC will reject it, costing you time and money.

In addition to COLEMS, there are some independent study guides and practice tests available. One such guide is available from TAB Books: "Practice Tests for Communications Licensing and Certification Examinations," by Sam Wilson and Joseph A. Risse. Persons requiring assistance in understanding some of the Radiotelegraph Elements or subelement topics (such as Antennas and Feed Lines/Element 3H) might contact a local amateur radio operator, packet radio operator or Radio Relay League member.

Do you, as an electronic repair technician need a Commercial Radio license to perform your duties? No, but there are many opportunities that await those who do hold such a license that are simply unavailable to those who do not. Many companies look to employ technicians who hold a commercial radio license at a higher rate of pay, as other shop technicians may work "under" the authority of a license holder (in some circumstances), meaning the license holder is ultimately responsible for the final inspection of the unit being repaired. Because of this, some shop owners will pay all fees for an employee to take a commercial radio licensing test; maybe the owner of the shop in which you work.

If you're interested in taking the test for the GROL or the GMDSS/M license, contact one of these agencies:

National Radio Examiners Division
The W5YI Group, Inc.
P.O. Box 565206
PO Box 565206
Dallas, TX 75356-5206
800-669-9594
817-461-6443
Fax: 817-548-9594
All elements are available on a monthly or quarterly basis, based on demand, at more than 250 test centers in all states.
Fee: $35.00 per license

Drake Training and Technologies
8800 Queen Avenue South
Bloomington, MN 55431
800-401 -EXAM
Fax: 612-921-7248

All elements are available on a daily basis at over 200 locations in all states except Maine and at over 300 locations worldwide. Evening, weekend, and holiday appointments are available.
Fee:$60.00 per examination
Contact: Julie Johnson

Electronic Technicians Association

International, Inc. (ETAI)
602 North Jackson Street
Greencastle, IN 46135
317-653-4301
317-653-8262
Fax: 317-653-8262

All elements are available at test sites throughout all states. Also at stateside
and overseas U.S. military installations (DANTES). Call for schedule
information.
Fee: $35.00 to $75 00
Contact: Anne Voiles

Elkins Institute, Inc.
P.O. Box 797666
Dallas, TX 75379
800-944-1603
Fax: 214-732-0244

All written examinations are available at test sites throughout all states.
Scheduled and "by appointment" examinations are available.
Fee: $50.00 for first element $25.00 for each additional element taken at
same sitting.
Contact: Ed Lyda

International Society of Certified
Electronics Technicians (ISCET)

2707 West Berry Street
Fort Worth, TX 76109
817-921-9101
Fax: 817-921-3741

All elements are available by appointment from 360 examiners in 47 states,
Guam, and some foreign countries. Examinations are not available in Alaska,
Vermont and Wyoming.
Fee: $25.00 to $75.00 per element
Contact: Dept. 19
Contact: FCC Technician Testing Center
800-759-0300
Fax:(703) 836-1608

National Association of Business and Educational Radio, Inc.
(NABER)
1501 Duke Street
Alexandria, VA 22314
Registration: 800-869-1100
Fax: 612-832-1290

Written elements 1, 3, 7, and 9 are available at 95 test centers nationwide
five days a week.
Fee: $63.00 to $120.00

SeaSchool
59054th Street N.
St. Petersburg, FL 33703
800-237-8663
Fax: 813-522-3155

All elements are available by appointment in 83 coastal cities
Fee: $25.00 - $55.00
Contact: Len Wahl

Sylvan KEE Systems
9135 Guilford Road
Columbia, MD 21046
800-967-1100
Fax: 410-880-8714

All elements are available seven days a week, walk-in or scheduled appointment (except holidays) at over I IO computerized testing centers in 35 states
Fee: $50.00 to $75.00
Contact: National Registration Center

The National Association of Radio Telecommunications Engineers, Inc.

NARTE
PO Box 678
Medway, MA 02053
508-533-8333
Fax: 508-533-3815

All elements available by appointment quarterly at NARTE test centers at 120 US universities and colleges. Also available at US and some overseas military bases (DANTES)
Fee: $40.00 per examination per sitting

Chapter 17

What Do You Know About Electronics? More on the AM Radio Detector For IR Remotes

By Sam Wilson, CET

Here is some very valuable information from Paul R. Dedrick of North Carolina. He explains why the infrared remote control signal for consumer products can be picked up by an AM receiver.

Dear Sam: I am writing in response to your "What Do You Know About Electronics?" column in the July 1992 issue of ES&T. Until recently, I was employed as a Technical Writer/Trainer. I taught classes in all consumer products to authorized repair facility technicians. What follows is the reasoning behind the AM radio test to detect the presence of infrared remote control signals. An additional test will be given. To begin, most infrared remotes operate by using a counted down fundamental frequency to generate a clock for data pulses which provide the different functions for the product to operate remotely.

What is this fundamental frequency? Well, it usually is (you guessed it) 455KHz. Most remotes use a ceramic resonator to generate this frequency. Often this delicate resonator has very thin leads which, when the remote is abused (i.e., dropped) the ceramic resonator becomes detached from the circuit. This can be determined by shaking the remote unit gently. If you hear a rattling sound, the resonator is usually loose! In this case, the remote can generally be repaired using a universal 455KHz resonator available from most Radio Shack stores.

If the resonator is not loose inside the remote, it may still be damaged, so you check it by operating the remote in close proximity to an AM radio. This tells if the remote is operating on frequency, by generating the buzz in the radio speaker.

To fully test the remote for proper operation, use an infrared detector card, available from most mail-order parts sources, manufacturers, Sencore, or Radio Shack. If the remote is on frequency, and putting out infrared pulses, there is a 99% chance that it is a good remote!

Also, I would be interested in receiving an author package for writing articles for your fine publication."

Seven years ago, as a high school electronics teacher, I used articles from ES&T and your TYEK tests to enrich my curriculum in my classes. The magazine is an invaluable aid to the continuing educational needs of the industry. I find it very informative and helpful in my efforts to keep current.

Figure 17-1 shows a schematic diagram from one of the training manuals I wrote, which shows a typical remote control transmitter construction. The transmitter is an infrared type, which offers up to 32 remote control functions. The transmitted signal is composed of four different codes, Custom Code, Inverse Custom Code, Data Code, and Inverse Data Code. Each code consists of eight bits. One single transmission consists of a combination of 32 bits. Besides the four codes, there is a leader code included in the transmission signal.

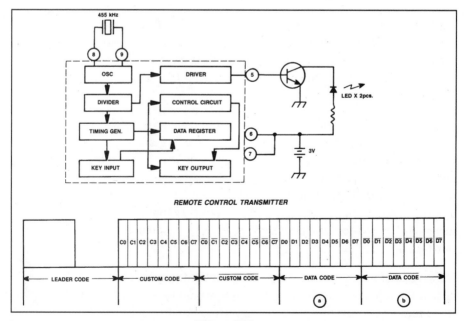

Figure 17-1.

It is made up of a 9ms carrier wave and 4.5ms off-wave. This is transmitted prior to the other codes. The Leader code is used to allow the microcomputer to differentiate the remote control signal from other control signals in terms of the time relation among them. The other four codes that follow the leader are applied to the microcomputer, which reduces them each to a 1 or 0 pulse. This is performed in accordance with PPM (Pulse Position Modulation) system. The pulses "a" and "b" are used in 32 different combinations of 1 and 0.

Sincerely, Paul R. Dedrick, CE
Secretary, North Carolina Electronics Association

Sam Says - Observe from this information that the signal from the IR (infra-red) remote control is pulsed. That is what makes it possible to inject the i-f signal into the AM radio and hear the signal in the radio speaker. A pulsed signal is rich in odd harmonics. As you know, a 455kHz sinewave signal could not produce any output sound in the receiver since it is the job of the "second" detector to remove a 455kHz carrier. The AM radio signal tells you that the pulses are being generated but it doesn't tell you anything about the condition of the infrared LED. The test with the infrared detector card suggested by Mr. Dedrick sounds like it will give more reliable information. I will be glad to consider any additional information on troubleshooting by our readers. Many, many thanks to Mr. Dedrick for the letter.

Circuits For Building the Microprocessor

In the last issue I said I was going to give an experiment using an off-the-shelf memory. I don't quite get that far in this issue. The reason is that there are some circuits external to the memory that must be built before we can use it. Some of those circuits are given in this issue. In the next issue the memory circuit will be assembled.

Keep in mind where we are going with this series of experiments. My contention has been that a microprocessor and a computer are both memory controllers. That is why I have spent some space in previous issues on the most popular memories used in μP and computer systems. Starting with this issue I am going to give a series of experiments in which the major circuits in a microprocessor are built on plug-in boards. We will perform the same operations as the μP would do to get the same result.

Many technicians have told me they can best understand theory by hands-on work with devices. I have to admit I do not understand how that works. It doesn't work for me.

I have spent a lot of time doing hands-on experiments that didn't give me my time's worth. In other words, there was a lot of constructing, measuring and troubleshooting just to "learn" one single piece of information that I could have read in a half page of typed material. I just don't understand trading 6 hours of wiring and pushing buttons for a minute of reading time.

Well, that's my concept. I know from experience that something can be true whether I understand it or not. So, to give those technicians the hands-on they need, I give μP experiments.

When a microprocessor is built with individual integrated circuits instead of on a single integrated circuit chip it is called a bit slice. So, basically, we are going to build a bit slice on plug-in boards. However, when we get all of the circuits assembled we will NOT be able to hook all of them together to get the bit slice.

The reason you can't put them together is that the timing of the various operations is very critical. So critical, in fact, that a microsecond difference in the arrival of pulses can make the system fail. However, when we have finished with the experiments we will have built the complete bit slice. If you are not into hands-on experimenting, read the experiments anyway. The basics of μP operation are explained in the theory writeups.

Memory Experiment

To demonstrate how the μP operates a Random Access Memory (RAM) you will write a telephone number into the memory, then, read the memory to get the number back.

I chose CMOS integrated circuits because I thought it would be convenient for the reader to operate the circuits with a 9V battery. However, the memory I chose got hotter than a $2.00 pistol on 9V, so, I changed to a +5V regulated supply. If you don't have a regulated +5V, build the one shown in *Figure 17-2*. It is useful for many other things besides these experiments.

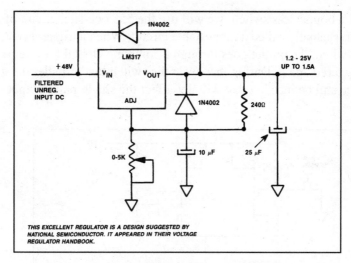

Figure 17-2.

Figure 17-3 shows three circuits to be built to engage the memory.

Figure 17-3(a) shows the circuit for our clock generator. It produces the timing pulses for all of the µP experiments. Computers are often evaluated by their clock frequency. Ours will generate about one clock cycle each second, usually written as 1Hz. That is somewhat slower that the 25MHz frequency of some desk-top computers.

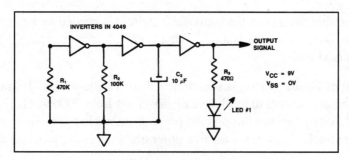

Figure 17-3a.

Mechanical switches and relay contacts have a habit of bouncing a few times when they are closed. Those bounces are interpreted by logic circuitry as being combinations of ones and zeros. That really messes up the operation. To get around that problem there are two circuits available to us. One is the bounceless switch (not shown). It is made with cross-coupled gates.

Instead of a bounceless switch, we will use the 555 one-shot circuit of *Figure 17-3(b)*. It is usually called a monostable circuit. When a trigger is received from the switch, the output goes through a complete ON-OFF cycle before it can be triggered again. During that cycle the switch that provides the trigger can bounce and bounce but that will not affect the single-pulse output of the 555.

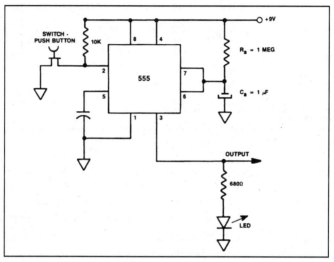

Figure 17-3b.

The monostable circuit (or, the bounceless switch) is needed so we can operate the system one step at a time. That is called single-stepping. More on that in the next issue.

The circuit of *Figure 17-3(c)* is sometime called "divide-by-16." It can be used as a binary counter to produce a binary count from 0000 to 1111. It will be needed to step our way through the program stored in our memory rows and to keep track of where we are in a program. So, it is our program counter.

Remember, the circuits of *Figure 17-3* are used by the μP to operate the memory. To test the circuits, first connect the output of the monostable circuit to the input of the counter (CL I of the first flip flop). Each time you momentarily switch the trigger input of the 555 monostable you get one output pulse. The output is shown by LED #1. It should be on for a short period of time after you operate the switch.

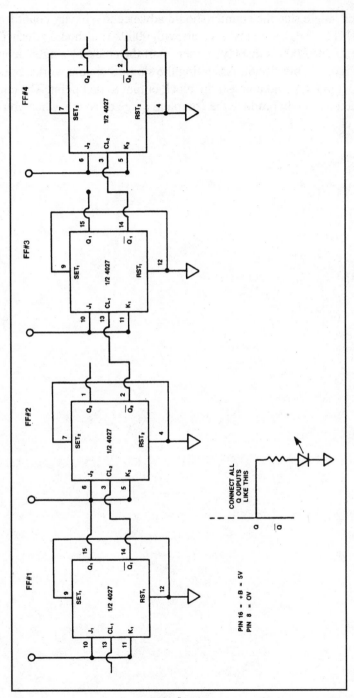

Figure 17-3c.

With each single step the counter should advance to a binary count from 0000 to 1111. The count may not start with 0000. You should advance the counts until the 0000 is displayed, then, start the count by single-stepping with the monostable circuit. Assuming the single stepping works, remove the 555 output to CL 1 and connect the clock output to that point. The counter should automatically produce the complete range of counts when you single-step the clock pulses.

Chapter 18

Lightning Protection For Audio Gear

By John Shepler

The thunderstorm season is upon us. This is often a particularly viscious time of year for damage to electronic devices caused by lightning. Let's review some of the measures we can take to protect valuable gear, and why lightning isn't always the culprit.

Most of us are careful not to get caught in the open during a thunderstorm. We've all heard the stories of people who have become human lightning rods or made the mistake of standing under the tallest tree on the golf course. Likewise, we assume that it's the equipment hooked to tall towers or sitting out in the open that is most vulnerable. Unfortunately, that is a false assumption.

A lot of equipment damaged by thunderstorms is sitting high and dry on living room shelves or wooden service benches. Many test setups and audio gear, other than receivers, don't even have antenna connections. Yet, a violent storm can race through your area and destroy a room full of equipment in a matter of minutes.

As you probably have guessed, the lightning strike isn't coming through the air in the building to do its damage. It has a much easier path through the power wiring. What happens is that lightning will hit an exposed nearby power line and cause an instantaneous voltage surge.

That surge has a fairly fast rise time and may persist for tens of milliseconds before fading out. Otherwise it looks like any other power line voltage. This means that the surge will pass right through transformers, along underground wires, into any building, and through the wiring and outlets into vulnerable equipment.The high voltage surge exceeds the ratings of semiconductors only for an instant, but it only takes microseconds to destroy solid state diodes, transistors, and integrated circuits.

Rural areas seem to get hit harder than cities. Storms at night also seem to do more damage. Most likely, this is due to the lower loading of the power lines at night and in the country. All those industrial motors and electric lights in the city help to absorb the excess energy caused by a lightning strike. Where there are fewer loads, the remaining equipment sees a larger transient.

This suggests some easy protection methods. Turn off electronic gear not needed, especially during a storm. The line switch will break the surge path for most equipment. Better yet, unplug anything particularly sensitive ... especially anything remotely controlled that is always powered. High priced televisions, stereo receivers, and personal computers are too valuable to risk operating during a storm. Also note that computers tend to crash during thunderstorms anyway, due to momentary power outages.

A good safety feature is the surge protected outlet strip. Most of these use a device called a MOV or Metal Oxide Varistor. This is a semiconductor that acts like back to back Zener diodes connected across the power terminals. The MOV draws little current until its rated voltage is exceeded. At that point the device resistance instantly drops and conducts current until the voltage goes below the lower threshold. MOVs work by "shorting" the transient before it gets farther into the equipment.

There are other semiconductor devices that provide the same action, perhaps clamping a bit faster. When the surge absorbing components are combined with coil and capacitor filters, you have a line protector that will also help filter out RFI and other hash on the power lines.

Surge protectors are also a good investment because not all power line transients are caused by lightning. When power companies switch from one feeder to another, lightning-like voltage surges are created on the lines.

Don't forget those antenna inputs. Cable TV lines and any outdoor antenna should have an antenna discharge unit or grounding block connected to a grounding rod that is driven into the soil. Cable installers generally provide these right where the line enters the house. People installing their own TV, FM, or scanner antennas may neglect this important connection. While the equipment will work just fine normally, blown RF preamplifier stages can result when a big storm comes overhead and causes high voltage static charges to build up on the antenna.

Now the most important point. Besides protecting the equipment, installing antenna discharge units, line surge protectors, and grounded outlets can help keep you and your customers alive. Now, that is well worth the cost and effort.

Chapter 19

Why Radio Sometimes Sounds Inferior To CD or Tape

By John Shepler

One disappointment common to many who buy expensive stereo systems is that the radio section often doesn't sound nearly as good as the tapes or CDs. You might be tempted to write this off to limitations in tuner design or radio wave propagation. For AM signals that is often true. For FM, however, the major limitation in sound quality occurs within the radio station's equipment.

You probably know that the FCC has reasonably strict standards for FM stereo audio. Frequency response from 50Hz to 15,000Hz and separation to 30dB may not sound state-of-the-art, but if these were the only limitations, most listeners would be hard pressed to find a problem.

Surprisingly, technology is not the culprit. The major limitations to radio audio quality have to do with business decisions.

Extreme business pressure to reduce expenses encourages some stations to skimp as much as possible on maintenance, testing, and equipment upgrades. While solid state control boards and newer transmitters can be ignored for years with little negative impact, cartridge and reel tape equipment, the mainstay of many stations for commercials and music, need more frequent attention.

Tape recorders and players are notorious for drifting out of alignment. Without regular tweaking to NAB (National Association of Broadcasters) standards, this equipment will gradually sound duller and duller until repairs are finally made.

CDs, on the other hand, either sound great or don't play at all. Compact disc is helping radio stations to sound consistently better, but is only starting to be practical for local recording. Digital tape is certainly an improvement over analog, but the recorders and tapes are still somewhat expensive.

The other major influence on station sound quality is the competitive pressure to stand out on the dial. The general public may be under the impression that a radio station is simply a wireless link from a CD player to their home systems. Nothing could be further from the truth. Most stations use extensive audio processing prior to transmissions. This processing is in the form of gain riding, equalization, compression, limiting, and even clipping. The purpose of the processing is to make the station sound loud and distinctive, so that listeners will tune no further.

Audio processing can be made to sound nearly transparent or very irritating, depending on how much is applied and how clean the audio is to begin with. The minimal amount is just enough peak limiting to prevent overmodulation of the transmitter. Most stations go well beyond this. If they didn't, their signals would sound weaker and presumably attract fewer listeners. Unfortunately, it is easy to adjust theaudio processors incorrectly and overdo the desired effects.

If you listen carefully to each station in a given area, they all sound a bit different. Some have wide stereo separation. Others have little stereo effect. Some sound very clear and distinct. On others, it is hard to pick out the instruments in a song. Some announcers sound clear. Others sound like the microphone is distorting. The audio processing is probably what is causing many of these problems.

Knowing what is happening on the radio dial is useful to servicing technicians in a couple of ways. You can reassure customers that their receivers are functioning correctly and then help them find the higher quality broadcasts. Then, once you have educated your customers, maybe they will be motivated to voice their complaints to some of the offending radio stations. Perhaps when the stations become aware of how important the audio quality is to their listeners, they will make the necessary changes to improve it.

Chapter 20

The Digital Pot

By Vaughn D. Martin

Control of most audio circuits is still accomplished the same way it has been for the last fifty years. The control element is the mechanical potentiometer. From the volume control knob to the sliders on an equalizer, the control judge is a human-the feedback is through the ears. Microprocessors have entered nearly every other segment of electronics, including the audio segment, but they always are stopped by the mechanical potentiometer.

This article focuses on microprocessor control of conventional audio circuits through the use of digitally controlled potentiometers. However, these devices can be applied to many other applications as well.

Conventional Audio Control

Designs incorporating mechanical potentiometers are still found in the majority of audio applications. The volume control on most car stereos is a rotary potentiometer. Volume control circuits generally resemble *Figure 20-1*. In this design, the potentiometer is used to control the signal reaching a fixed gain amplifier section. A potentiometer in this application would likely have a logarithmic taper, since volume is a logarithmic function.

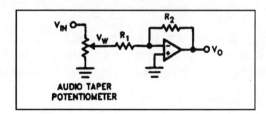

Figure 20-1. A conventional pot used as a volume control.

Tone controls can vary from single pot and capacitor circuits to complex active filters. The Baxandall filter network has been the workhorse of the audio industry for years. This design, illustrated in *Figure 20-2*, utilizes two linear taper potentiometers to control the gain of an active filter. In this configuration, the potentiometer replaces a portion of both the input and feedback resistors. By moving the position of the wiper, both resistors change in opposite directions.

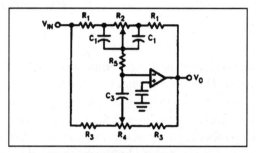

Figure 20-2. *A baxandall control.*

Graphic equalizers are one of the fastest growing modes of audio control. A graphic equalizer contains a group of bandpass filters, usually seven. Each filter has a potentiometer controlling the gain to that band pass. Potentiometers generally appear as sliders on the face of the equalizer.

A typical graphic equalizer schematic is shown in *Figure 20-3*. EQs are used to compensate for the imperfections of a listening environment by boosting or cutting gain at specific frequencies. By using a spectrum analyzer and a "pink" noise generator, the response of an audio system can be customized for a particular room or concert hall. This is accomplished by inputting a desired response to the system-generally flat across the audio band, with some attenuation at higher frequencies, often referred to as "pink" noise. The equalizer is then adjusted until the system output, displayed on the spectrum analyzer, closely matches the pink noise input.

This process of matching a system to a room is often referred to as environmental calibration. It is a process requiring the listener to read the display of the spectrum analyzer and manually adjust the potentiometer/sliders of the equalizer.

The heart of the control of each of the circuits described earlier is the mechanical potentiometer. Automated control of these devices is a challenge. Clearly, microprocessor control of these functions is desirable. The control elements utilized for automated control are discussed below.

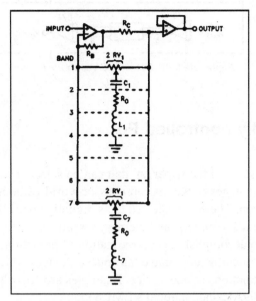

Figure 20-3. A graphic equalizer configuration.

Automated Control Elements

While these devices are primarily used for industrial control applications, motorized potentiometers offer a relatively straightforward approach to simple audio control circuits. In these devices, a dc reference voltage, or a digital signal representing position is input to a small motor assembly that is linked to a rotary potentiometer. Drawbacks to this type of system are numerous, including noise caused by the motor assembly as well as the increased space and power requirements of placing a motor on an audio PC board.

D/A converters can also be used to control and manipulate analog circuit functions, but introduce more complexity. These devices are the choice of high fidelity digital audio controls due to their high precision. But for the analog circuit designer, they can be a little intimidating. For example, one way to control volume with D/A converters is illustrated in *Figure 20-4*. In this circuit, the signal is sampled with an A/D converter, manipulated by a microprocessor, and returned to the analog world with a D/A converter. This design entails sampling, real-time processing, as well as A/D and D/A conversions. Not only may the analog designer be faced with portions of his circuit that may be unfamiliar, the results may be overkill.

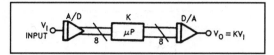

Figure 20-4. *An A/D and D/A controlled volume control.*

The Digitally-controlled Pot

An array of resistors with a wiper tap that can be selected with digital control offers many advantages of the microprocessor world without the complexity of D/A conversion. These are referred to as digitally controlled potentiometers. Logic circuits, counters, and memory circuits are often teamed up with resistor arrays to accomplish an approximation of potentiometer control. Recently, a few manufacturers have introduced devices which incorporate many of these functions in one device. Examples are Xicor's X9MME, Toshiba's T09169AP, and National's LMC835.

The Toshiba and National parts are designed around specific audio applications and are distinctively different from the Xicor device. They incorporate features that lend themselves well to audio designs, but are not intended for general purpose potentiometer replacement. Moreover, they offer only a limited number of wiper positions.

Xicor's X9MME combines a single 99 position potentiometer with three line digital controls. *Figure 20-5* contains a functional diagram, pin description and mode selection for the device. In addition to the internal counter circuitry for wiper position control, this part also incorporates nonvolatile memory to retain wiper position. It has been designed as a digitally controlled replacement for the mechanical potentiometer.With its conventional three terminal potentiometer design, it integrates easily into existing analog designs.

To illustrate digital control of potentiometer circuits, the X9MME from Xicor was used to replace mechanical potentiometers in a well known audio circuit. The following should demonstrate the ease of designing with the X9MME as well as the advantages of microprocessor control in audio circuits.

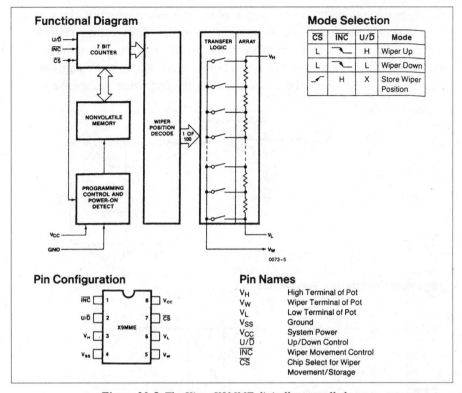

Figure 20-5. The Xicor X9MME digitally controlled pot.

The X9MME in An Audio Circuit

The Baxandall tone control circuit is the basis for the designs shown here. The following sections will discuss the principles behind the Baxandall circuit and then walk through the design utilizing the X9MME. Special design considerations for the X9MME will be discussed, and the performance and operation will be evaluated.

The Baxandall circuit, its response, and equations for gains and filter frequencies are shown in *Figure 20-6*. This circuit contains two active filters whose gain is controlled by two potentiometers. *Figure 20-7* illustrates the bass portion of the circuit. The maximum gain of this circuit is at low frequencies, where the capacitors in the circuit can be considered to be open circuits. The capacitors have been omitted for clarity. (The treble portion of the circuit, not illustrated here, follows along similar lines.)With the addition of another potentiometer on the output of the Baxandall network, the system represents a single channel of an audio preamplifier. The circuit contains three potentiometers which control volume, treble and bass. These pots would appear as knobs on the face of a home or car stereo, to be adjusted by hand to control and shape the sound reaching the amplifier and speakers.

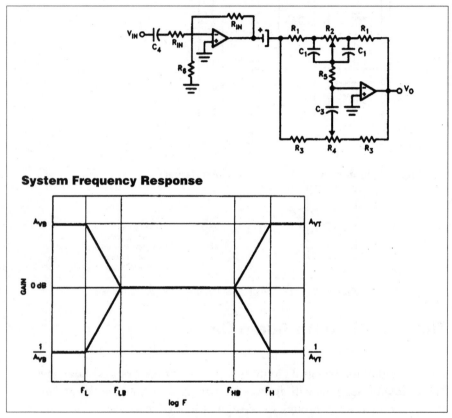

System Frequency Response

Figure 20-6. An active filter preamplifier.

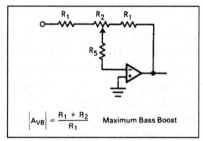

$$\left| A_{VB} \right| = \frac{R_1 + R_2}{R_1}$$ Maximum Bass Boost

Figure 20-7. The bass portion of the active preamp circuit.

Neglecting the digital control lines and 5V power for the X9MME, the circuit is shown in *Figure 20-8*. The X9MME will replace bass, treble and volume potentiometers. Note that this does not alter analog design considerations.

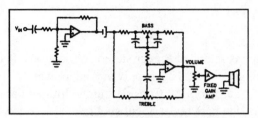

Figure 20-8. *An active preamp with bass, treble, and volume control.*

R_2 and R_4 are both linear taper pots. Since the X9MME is also a linear taper pot, it is a direct replacement. RV , the volume potentiometer, is specified as an audio taper pot, since it is used for volume control. By placing a small resistor from wiper to low on any linear pot, as shown in *Figure 20-9*, an audio taper can be approximated. In this case a resistor of one-tenth the total pot resistance is a close approximation of an audio pot (reference 1).

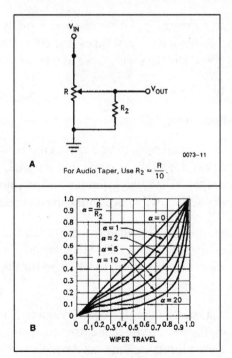

Figure 20-9. *Trimming a pot with an external resistor.*

This circuit is designed to have a gain of one across the entire audio range, with the potential for a boost or cut of 20 dB at the frequencies selected by the designer.

The Design

The design chosen is intended for car stereo applications. It should therefore operate from a single ended, 12V supply and adapt well to speakers that are commonly used in automobiles. Considering the limited bass response of most car speakers, the bass boost or cut should not be so low that the speakers cannot reproduce the sound.

The desired circuit would operate from a 12V power supply, have a 20dB boost or cut at 100Hz (bass) and 10kHz (treble). The available resistor values for the X9MME are 10K, 50K, and 100K.

Steps in the design:

1. R_2= 50 kilohm (arbitrary, X9503)

The design must start somewhere. This value was actually determined after running through the design a couple of times and comparing the values determined for the potentiometers with those available.

2. $A_{VB} = I + R_1/R_2$; for 20 dB (10),

$R_1 = R_2/9 = 5.6$ kilohm

The bass portion of the circuit must have a maximum boost of 20dB. This is determined with the bass pot all the way to the input side. A quick look at *Figure 20-8* illustrates this. Here, the formulas for the cutoff frequencies of the active filters are broken down to determine the element values to use.

Here, the maximum treble gain is calculated in similar fashion to the maximum bass gain. The circuit with the X9MME inserted is shown in *Figure 20-10*. These are the values that were used in lab experiments and for demonstration purposes.

The X9MME can be a source of high frequency noise. There are internal voltage generators on the device which are used to operate switches internally as well as to store information into the device's nonvolatile memory.

The principal noise frequencies begin at approximately 150kHz, and while this is beyond the audio range, it can still be a source of problems in the circuit. Capacitors were added around the X9MME to filter noise. These are included in *Figure 20-10*.

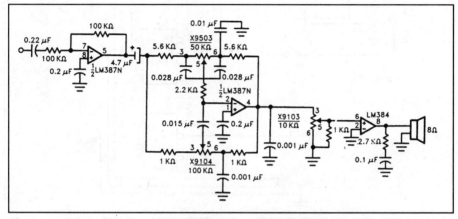

Figure 20-10. *A preamp with three digital pots (No digital controls shown).*

Digital Control

The digital control lines of the X9MME are $\overline{INC}$, $\overline{CS}$, and $\overline{U/D}$. $\overline{CS}$ (chip select) allows the wiper to be moved. $\overline{U/D}$ (Up/Down) determines the direction in which the wiper will move, the $\overline{INC}$ (increment) initiates movement on its falling edge. $\overline{CS}$ is also used to store the wiper position in non-volatile memory. When $\overline{CS}$ is returned high, a store operation is commenced.

When initially designing with the part, it was helpful to assemble a simple switch system for controlling the parts. A 555 timer was used to generate a fairly slow clock pulse and connected through a momentary switch to the increment pin of each X9MME. With pull up resistors on each digital line, a grounding switch was connected to $\overline{U/D}$ and another to $\overline{CS}$. To move the wiper up, $\overline{CS}$ was set to ground, $\overline{U/D}$ to 5V and $\overline{INC}$ pulsed with the clock. Each step of the clock produced a 1% change in wiper position. *Figure 20-11* illustrates the switching network that was utilized for controlling all three X9MMEs.

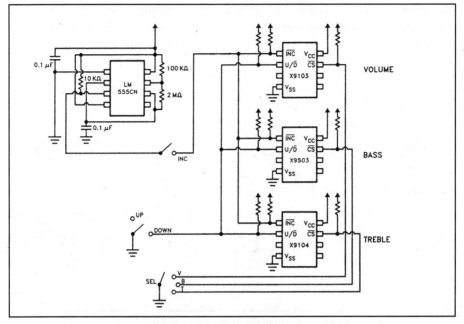

Figure 20-11. Switch network for manual operation.

This initial procedure allowed the analog portion of the design to be separated from the digital. Once the circuit was functioning adequately with the switch network controlling the X9MMEs, microprocessor interface was relatively simple.

Microprocessor Interface

With three devices on the board, 9 control lines are required. To simplify interface to an 8 bit microprocessor, the INC lines for all three parts were connected to the same pin. The pin configuration used for interface to the 6502 microprocessor system is as follows:

1 = Volume

2 = Bass

3 = Treble

To move the wiper of a given pot, that pot's $\overline{CS}$ is brought low, the $\overline{U/D}$ for the appropriate pot is asserted H or L depending on the direction of wiper movement, and $\overline{INC}$ is toggled. For example, to increase the volume the following two patterns are alternated to the port connected to the E^2 PREAMP.

$\overline{NC}$	$\overline{INC}$	$\overline{CS}$	$\overline{U/D}$	$\overline{CS}$	$\overline{U/D}$	$\overline{CS}$	$\overline{U/D}$
1	0	0	1	1	1	1	1
1	1	1	0	1	1	1	1

Note that CS has been selected, U/D set to 1 and INC toggled. Bass and treble settings are altered in a similar manner. The microprocessor system used in the lab consists of a 6502 based keyboard monitor. The controlling program scans the keyboard for a recognized ASCII character which transfers control to the specified subroutine. For any given input, the appropriate increment is toggled 10 times before returning to the controlling program.

An example of a volume, bass, or treble adjusting program, in the microprocessor's mnemonic code, follows:

```
        LDX   #00          Load counter with zero
0333    LDA   0006         Load accumulator with first pattern
        STA   A000         Output pattern
        JSR   ED2C         5 ms wait
        LDA   0007         Load 2nd pattern
        STA   A000
        JSR   ED2C
        INX
        CPX   0008         Compare counter to I0
        BNE   0333
        RTS
```

In addition to the adjustment subroutines, an initialization subroutine can also be called up. This subroutine sets the volume to zero and bass and treble to 50%. This is used to reset the controls. It would be used only during installation of the system.

This first section of the one time initialization program sets all pots to zero.

	LDX	#00	Load counter with zero
0111	LDA	0000	Load accumulator with first pattern (80h)
	STA	A000	Output pattern
	JSR	ED2C	5 ms wait
	LDA	0001	Load 2nd pattern (C0h)
	STA	A000	
	JSR	ED2C	
	INX		
	CPX	0008	Compare counter to 100
	BNE	0111	

This section sets the bass and treble pots to 50% and returns control to the controlling routine.

	LDX	00	Load counter with zero
012C	LDA	0003	Load accumulator with first pattern (85h)
	STA	A000	Output pattern
	JSR	ED2C	5 ms wait
	LDA	0004	Load 2nd pattern (F5h)
	STA	A000	
	JSR	ED2C	
	INX		
	CPX	0005	Compare counter to 50
	BNE	0333	
	RTS		

Operation and Performance

The E^2 preamp circuit operates much like many sophisticated home stereo systems today. All controls are digital switches-in this case, a keyboard for demonstration purposes only. There are no moving parts beyond the switches, and the entire system is relatively free from problems with vibration or jarring (potential hazards in mechanical pot systems).

Keys 1 through 6 on the keyboard represent the up down controls for the circuit. By depressing 1, the volume is increased by 10 steps. Key 2 decreases volume in the same way; 3 is treble up; 4 is treble down; 5 is bass up; 6 is bass down. The 1 key calls the initialization routine. Beyond allowing control of step size and the auto zero or initialize function, the present system does not take advantage of the versatility of microprocessor control.

Performance of the system was nearly identical to the same circuit with mechanical potentiometers. The X9MME is quiet to -65 dB below a 1V signal, which is fair for audio quality devices. For audiophile quality, this number should be around -120 dB, but in car stereo or communication equipment applications this device works adequately.

Aside from the obvious advantage of a smaller number of moving parts, the ability to choose step size in adjusting the controls has shown to be the most useful added feature. Ten steps per adjustment proved to be an easy value with which to work.

Having demonstrated the ability of the X9MME to replace mechanical potentiometers in analog circuits, more complex circuits may now be considered. With microprocessor control, advanced circuit design and digital control simply becomes an extension of the principles discussed so far.

Microprocessor control of this and other analog circuits is simple when utilizing a digitally controllable potentiometer. The gain of the entire circuit, or the boost or cut of a given frequency range is instantly alterable via microprocessor commands. Once control is assumed by the microprocessor, any parameter of the analog circuit that is controllable by a potentiometer is available to the programmer. For example, the graphic equalizer/spectrum analyzer combination discussed earlier can easily be automated once microprocessor control is assumed. By controlling the position of potentiometers that control the gain of the individual equalizer bands, the system frequency response can be calibrated to any room or listening environment.

Here is just one scenario: A "Calibration" button is depressed on the equalizing circuit. This activates a "pink" noise generator which sends a short burst of sound to the system. The spectrum analyzer in the system then decides which frequencies require adjustment, changes the positions on the appropriate potentiometers, and the system is calibrated. No sliders need to be adjusted; no separate (and expensive) spectrum analyzer; moreover, a relatively unsophisticated user can now perform an accurate environmental calibration of the system. A simpler version of an auto calibration circuit could be incorporated into home and car stereos as a one time only installation adjustment. When a car stereo is first installed, the installer would push the calibration button on the back of the unit. This would adjust a compensation circuit, separate from the main tone controls. The settings would then remain in the non-volatile memory of the digital pots until the system was upgraded or installed into another car. Thus the same unit would be customized for different speakers, different amplifiers, and even different auto interiors.

Chapter 21

Servicing Personal Headphone Stereos

By Sheldon Fingerman

In today's world of throwaway consumer electronics, servicing personal headphone stereos may seem like a waste of time. On the other hand, servicing these units can be a rewarding break from the routine, and will enable you to offer a service that is difficult to find. This helps to bring in customers. Even most authorized service centers will refer these miniature marvels to the factory.

Once inside a personal stereo, you'll find that there is very little difference between it and any other electromechanical device, except for size. Typical problems you'll find in these units are broken belts, cracked circuit boards, cold solder joints, and worn out parts. Approach problems in personal stereos the same as you would approach problems in any other consumer electronics product. The primary difference you'll find is that things are on a smaller scale.

Also, keep in mind that personal stereos suffer from abuses that in-home stereo products are not exposed to. When a headphone stereo is turned over to you for service, assume that it has been dropped, that the unit went one way and the headphones went the other, that it has sand in it, and that it may have been under water at some time.

Some Common Problems

The most common symptoms encountered in servicing of personal headphone stereos are intermittent audio, improper functions, or failure to operate entirely.

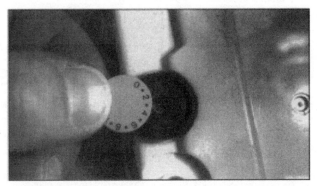

Figure 21-1. *Most screws on water resistant models can be found under adhesive "dots." Note the false hole to the right, and the dot that is still in place at the lower right.*

Intermittent audio is without a doubt the number one complaint with Walkman type personal stereos. Wiggling the headphone plug will usually make and break the signal. The customer will probably tell you that she has already purchased a new set of headphones but that the problem persists.

The only test equipment required to troubleshoot this problem is a known-good set of headphones. If the unit operates properly with the known-good headphones, you know that the problem is in the headphones. If the unit continues to exhibit the problem, the problem is in the stereo unit, not the headphones.

If the problem is with the headphones, and not the unit, a decision will have to made as to whether to attempt to repair the headphones, or to simply replace them. If the phones are valuable and the problem is one that can be repaired, it may make sense to service the headphones. If the headphones are of the $10 variety, replace them.

In most cases the problem will be in the unit itself, and is usually due to a broken solder joint, a broken foil, or a crack in the circuit board.

If the unit appears to be totally dead, the remedy may be as simple as replacing the batteries. As you are probably aware, just because batteries are new doesn't mean they are good. If the batteries check good, try cleaning the battery contacts. If you can hear the motor spinning when you hold he unit up to your ear the problem may be a broken belt. If you can't hear the motor spinning, the problem may be a broken belt. If you can't hear the motor spinning, the problem may be a dirty or broken leaf switch, or a circuit board problem.

Figure 21-2. *This latch is held in place by the outer cover (which is partially removed). It has a spring behind it, and will pop off if not held in place.*

Improper function can include just about anything. One common problem withauto-reverse units is hearing both sides of the tape playing at the same time, one going backwards. This is usually caused by a broken ground on the record/play head. Constant reversing can be cause by a broken belt. Also, look carefully for any broken or unhooked springs around the head, capstan and pinch roller assembly.

Problems Caused By Small Size

The very compact models with metal cases sometimes experience problems because of a bent case. There is very little space between the moving parts between the moving the parts and the outer case in these units, so just being packed in a tight suitcase may be enough to bring one of them to a complete halt.

Because many of the functions are driven by very small gears, the mere presence of a grain of sand can cause problems. If the mechanism seems to be jammed, ask if the unit has been taken to the beach recently.

Getting It Open

In most cases of problems in headphone stereos, the repair itself may actually be quite simple. The real problem in servicing headphone stereos is getting them apart.

Figure 21-3. *A transport with the circuit board removed.*
Note that the small cylindrical motor (lower right) is
turned 90 degrees to the capstan and flywheel.

There are two common methods of assembly. In the first method, the case is solid and there are no visible screws; the transport is held in place by screws inside the cassette well. The other is more typical, with external screws holding the outer cases on.

On some of the water resistant models the screws are recessed and covered by small plastic circular "dots" that match case color. Carefully lifting these dots will reveal the screws. Don't be concerned if you find no screws under some of the dots. Many of the cases are used on more than one model, and these dots may be covering nothing, or some of the holes may be for access to adjustments like speed, etc. any small, sharp, awl type tool will easily remove these covers. They have adhesive on one side, and I usually stick them to the edge of the container I use to hold the screws.

Be careful when removing screws that you assume are holding covers on. Slowly loosen each screw, and if it doesn't begin to back out right away it may not be a case screw. You don't want to inadvertently unscrew any internal parts, turning simple repair into a nightmare.

In some cases the cassette door will have to be removed, and the hinge released. The door may be held by two screws, one at each side, and the hinge may be held by a plastic retainer that resembles a tiny washer. These retainers can be removed with a jeweler's screwdriver. Get the flat blade under the retainer and twist, lifting it off. Try to keep a finger on some tape on it when you do this. When they come off they have tendency to fly across the room and you will never find them. Don't ask how I know this. If you find that you enjoy doing these repairs, it usually pays to keep some spare retainers on hand.

If you feel you've got all the screws out, but the cover still won't come off, it's probably being held on by small indentations in the plastic, in much the same way as a remote control is held together. Some careful prying, pushing, probing, and pulling may do the trick.

On water resistant types you will probably have to lift the assembly out of the case. Starting with the side opposite the function buttons, pull it up and away from rubber button caps, which will stay with the case. If you are working on the type with case covers, lift the side opposite headphone jacks, and then pull the cover away from the jacks.

When the cover starts to come off, use caution with regard to all external switchers. Some are held in place by the cover, others may stay on the chassis. One type of Sony model has a small spring loaded latch that keeps the cassette door closed. This is held in place by the cover and may fall out, springs flying across the room never to be found. Again, don't ask how I know this.

Figure 21-4. The circuit board around the headphone jack(s).
The pointer shows one of the screws holding the board in place.
Note the arrow pointing the screw as well (Sony).

Figure 21-5. This knob and switch must be lined up on reassembly. Note that headphone jack (in the case with wires attached) is not part of the circuit board on this model.

Some units have battery compartment doors (spring loaded in some cases) inside the cassette compartment. These too may be freed when the transport and case are separated. A little strategically placed tape will keep these parts where they belong until reassembly.

Your goal is to try to disassemble the unit just enough to do the repair. It may mean removing the entire transport and circuit board assembly, or merely removing a cover. Every model is different. A few minutes of thought at each stage of disassembly may save you a lot of time later on.

Figure 21-6. This unit was dropped, causing the head assembly to jump behind the play switch on this autoreverse model. The cover wouldn't open all the way, and when I tried the reels started turning. Prying the switch back in place solved the problem.

Visual Inspection

Examine the area around the headphone jack for problems, and solder if necessary. Small cracks in the board and foil can be repaired with solder. In some cases you may want to bridge the crack with a piece of bare wire. This is the most common of all personal stereo repairs. If the case cover comes off easily, the circuit board is under the cover, and the foil side is up, you are probably looking at a five minute repair.

The main circuit board is usually under the tape transport, and should only be removed if necessary. It's held in place by screws, and, or, plastic tabs protruding through the board, possibly at the perimeter. You will probably notice lots of tape holding wires in place. It may have to be removed to gain access to screws, and to put a little slack in the wires. Pull the tape off carefully. Most of the wires are tack soldered to the circuit board; if you pull one free you may never figure out where it went.

Correcting Tape Transport Problems

If the problem is with the tape transport and you have to remove the circuit board to get to the transport, remove all screws (look for arrows on Sony products), and carefully lift the board out of the way. If the board won't budge, look for any components that may be soldered through the board.

If the problem is a broken belt you may have to experiment to get the new belt on correctly. Most modern transports have very small cylindrical motors mounted at an angle of 90 degrees to the flywheels, counterbalancers, and reels. This means you will have to put a half-twist in the belt to get it on right. Figuring out how to put on a new belt without a service manual can be very time consuming. Once you've replaced a belt, use the external power jack, clip leads, or even a battery to bring the transport to life to make sure things are turning in the right direction before putting everything back together. If you have to remove any flywheels or capstans, be sure to check for spacers that may have been used to replace them. Space is at a premium in these things, and one spacer, no matter how insignificant it may seem, may make the difference between two parts working in harmony, or agony.

Putting It Back Together

Assembly is exactly the reverse of disassembly. Take notes as you work that you can refer back to.

Make sure all wires are free of any moving parts, and retape them down if necessary. Some units have leaf switches and record/play switches that must be lined up between the transport and the circuit board. Miss one and you'll have to go back in again.

When replacing the cover, make sure any miniature knobs are lined up with their respective switches. This is an easy one to forget. A bit of tape, and an extra minute or two will save you a lot of grief. Also, be sure that you have not pinched any wires, and that any battery compartment covers are lined up properly.

If the cassette door was removed you have an opportunity to align the heads and clean the capstan(s) and pinch roller(s). If you have to replace the small retainer on the door assembly be careful. Try just pushing it on with your finger, or if that doesn't work use a small screwdriver. Think! If it slips, falls, or flies off where will it go? You certainly don't want to spend an afternoon crawling around looking for it, or worse, having to take the unit apart again because it fell inside.

Once the unit is back together, check all functions, and make sure the indicators on the volume, and if applicable, tuning dial are not turned 180 degrees.

Establishing Your Policy

Living at a ski resort gives me the opportunity to see more than my share of personal stereos. I charge a flat rate that is less that the factory charges. I also do not offer a warranty beyond the front door. I feel it's better to be up front about these policies than to have the same unit back in the shop every day because the customer keeps dropping it. After a repair they either work or they don't.

Also, working on personal stereos has to be approached as a public service that can bring in a few bucks. Avoid "bizarre" problems, and learn to say "No!" or at least, "I'll get to it when I can." Unless business has been particularly slow, give yourself a maximum amount of time to invest in each repair. If you find that you are having difficulty getting a cover off, give up and cut your losses quickly.

The real money maker is working on high-end microcassette recorders. They experience the same problems as personal stereos, but repairing them is like doing microsurgery. Because they cost around $200 when new, you can usually get your normal hourly fee, and offer a warranty. Be prepared to work with wafer thin circuit boards that contain micro leaf switches and surface mount components.

You may find, as I have, that servicing these items can be enjoyable and profitable, and in addition may generate business in other areas. As with other products you service, the quick jobs will make up for the headaches.

Chapter 22

Ten Important CD Waveforms
By Homer Davidson

When you're faced with a malfunctioning compact disc player, observation of some important waveforms in the various circuits of the unit may help to locate the defective circuit. A laser power meter test will determine if the laser diode is defective (*Figure 22-1*). Absence of RF or EFM waveforms may indicate a defective optical assembly or RF amplifier transistor or IC.

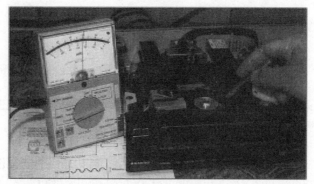

Figure 22-1. *Measurement of the infrared power output of the laser assembly using a laser power meter may uncover a defective laser diode assembly.*

The presence of a PLL-VCO waveform indicates that the VCO (voltage-controlled oscillator) circuits are functioning normally.

Presence of the focus error (FE) waveform, which is used in making focus tracking adjustments, across the focus coil may indicate that the focus circuits are normal.

A dead motor may produce a waveform that is a straight white line, while a rotating motor will produce a moving waveform. Waveforms taken on the sled and spindle or disc motors may indicate that all of the motor circuits are normal. Observation of waveforms within the compact disk circuits may help to locate the defective component.

Laser Power Meter Measurements

The best way to determine if the CD player's laser circuits are functioning is to measure the output of the laser with a laser power meter. Excessively high or unusually low power may indicate a defective laser diode. In most cases, if the laser diode assembly is defective the CD player will not be worth repairing.

Compare the laser power meter reading with laser power specification on the schematic or the CD player's power label. Check to be sure that the correct supply voltage is applied to the optical pickup assembly and observe the RF, or EFM, waveform to see if it looks correct before concluding that the laser head assembly is faulty.

Laser Power Adjustments

Laser power adjustments are made to insure proper operation of the laser diode. If the power output of the laser diode measured by the laser meter is near the value specified by the manufacturer's service data, most likely the laser diode is normal.

A low EFM waveform may indicate a defective diode. In some optical pickup assemblies, the VR adjustments mounted on the laser head assembly are adjusted at the factory and should not be touched. Often, laser power adjustment is needed only after replacing the laser diode optical assembly.

To check the laser power and make any necessary adjustments, place the power meter probe over the lens assembly and note the meter reading. Remember, never look directly at the laser lens assembly, and always keep your eyes at least a foot away from the laser when the unit is operating. With the interlock defeated, the laser diode will emit infrared light.

Locate the power adjustment (VR) on the main board or optical assembly. Check the manufacturer's test points and different control adjustments on the printed circuit board. Adjust VR so that the laser power output corresponds with the value specified in the service literature.

Test Points and Adjustments

Most CD players have test points and VR control adjustments on the main PC board. Often, however, the laser head adjustment is located on the optical assembly. Locate all test points before you try to take scope waveforms. Besides being useful in locating the cause of a defective circuit, the same test points may be used for making adjustments as well. Check the manufacturer's service literature for test points and VR adjustments.

Adjustments to the compact disc player should be made in an established sequence. Some manufacturers specify this sequence: laser power, PLL-VCO, focus offset, tracking offset and tracking gain. Other manufacturers list the required order of adjustments as laser power, skew, HF or RF level, focus gain, tracking gain, PLL-VCO, focus offset and tracking balance. Special screwdrivers or alignment tools may be needed for some of these adjustment screws. All adjustments are made with a test disc playing.

Waveform 1- RF or EFM Signal

A normal RF or EFM signal will indicate that the laser diode optical assembly and RF transistor or IC are functioning. The EFM waveform may be observed at pin 20 of the RF amp (IC101) (*Figure 22-2*). You may find a test point given in the manufacturer's service literature on the main PC board (TP-RF). If the EFM signal is not present, the player may search and shut down. Check the laser diode and RF circuits in the case of constant shut down of the CD chassis.

The normal EFM waveform should have a good eye pattern, represented by a clear-cut diamond outline. A vibrating or jittery eye pattern may indicate poor adjustment. Correct adjustment is to make the diamond-shaped area as clear as possible without excessive jitter (*Figure 22-3*).

To adjust the EFM signal, locate the RF, HF or EFM scope test points and level or gain control on the main PCB. Check the RF pattern on the scope. Adjust the tangential adjustment screw for maximum waveform or voltage reading as given in the service manual.

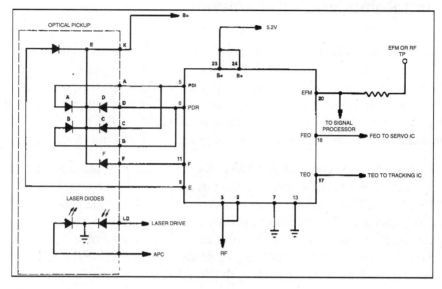

Figure 22-2. The EFM or eye pattern waveform may be observed at pin 20, or at the test point (TP-RF) in the CD player.

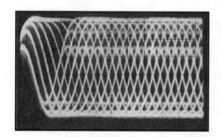

Figure 22-3. A good EFM waveform voltage should have a clear-cut diamond shape with very little jitter.

For example, in the RCA MCD-141 CD player, adjust the control so that the EFM signal waveform is 1.3Vpp on the oscilloscope. Similarly, in a Sanyo CP660 model, adjust the HF-EFM waveform to 1.5Vpp on the oscilloscope. A clean-cut and normal EFM waveform confirms that the laser pickup assembly and RF amp circuits are normal.

Waveform 2 - PLL-VCO Signal

Presence of the phase-locked loop voltage-controlled oscillator (PLL-VCO) waveform indicates that the EFM voltage has reached the digital signal processor and servo IC. The VCO oscillator output signal is compared to the edge of the EFM signal read from the disc. The PLL-VCO oscillator adjustment can be made with the oscilloscope and frequency counter.

Some manufacturers use a frequency counter to set the frequency of the PLL-VCO, while others make the adjustment on the oscilloscope (*Figure 22-4*). Adjustment of the PLL-VCO frequency must be correct so that the optical lens assembly follows the disc motor and responds to dropouts caused by defects on the surface of the disc.

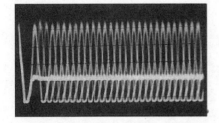

Figure 22-4. The PLL-VCO waveform voltage can be measured and adjusted from ground (bottom) to white or dark line. Some manufacturers use a frequency counter to adjust for the correct frequency.

Locate the coil or transformer on the PC board, and connect the oscilloscope to the correct test points. Play a test disc. Observe the waveform on the scope. Adjust the coil until the specified dc voltage is from ground to a white or dark area on the waveform (1.2V to 1.5V). In a similar manner, connect the frequency counter to the correct test point and adjust the coil for the correct frequency listed in the service manual. Follow the manufacturer's PLL-VCO adjustments.

Waveform 3 - Focus Error Signal (FE)

Presence of the focus error (FE) signal waveform indicates that the correct focus signal is reaching the RF amp to the focus servo IC (*Figure 22-5*). Adjust focus gain and offset for maximum output. The focus offset may include the RF and EFM adjustments, or they may be separate adjustments, depending on the manufacturer of the CD player.

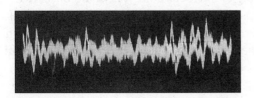

Figure 22-5. The focus error (FE) signal is taken at the RF amplifier IC or transistor circuit.

Again, depending on the manufacturer, the focus offset adjustment may be called the jitter or eye pattern adjustment. The focus offset adjustment is about the same as the jitter adjustment (*Figure 22-6*). This focus adjustment can be made with a standard music compact disc and an oscilloscope. Connect the scope to the circuit test points on the pc board and play the disc. Adjust the VR focus offset adjustment until the EFM or RF waveform is maximum, and displays a sharply defined diamond-shaped pattern.

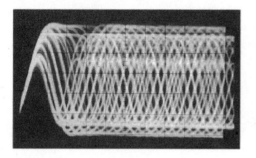

Figure 22-6. The jittery or bouncing EFM waveform may indicate improper focus error adjustment.

Waveform 4 - Focus Coil

If you're able to observe a focus coil waveform, you know that the focus signal is reaching the coil itself (*Figure 22-7*). The FE signal is sent from the focus-tracking servo to a focus driver IC or transistor that drives the focus coil located in the optical lens assembly.

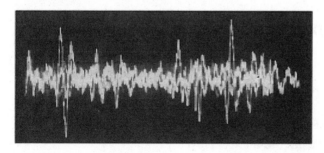

Figure 22-7. Connecting the oscilloscope across the focus coil winding indicates if signal is present at the coil assembly.

If the focus drive signal was missing at the focus coil, you would observe a horizontal white line on the oscilloscope at this point. The focus error signal can be checked at the FE output terminal of the focus-tracking servo IC, then followed to the focus driver and then observed at the focus coil leads (*Figure 22-8*). If you suspect that the focus coil may be open, turn power to the CD player off, then check continuity of the coil using the low ohms range of the ohmmeter.

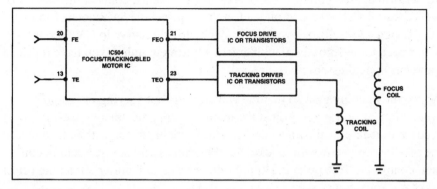

Figure 22-8. The focus error output (FEO) and tracking error output (TEO) signals are fed from the focus/tracking/sled IC.

Waveform 5 - Tracking Error Signal (TE)

If you are able to observe the tracking error (TE) signal waveform at the focus-tracking servo, you have confirmed that the tracking signal has reached this point. The TE input and output (TEO) signals may be traced with the scope (*Figure 22-9*).

Figure 22-9. The tracking error (TE) signal can be signal traced with the scope from tracking IC to tracking driver and tracking coil.

Usually, the tracking gain control is located at the TE input terminal of the IC. The presence of normal TE input and output signals indicates that the tracking signal is good to this point in the circuit.

When you observe a normal TE signal at the input terminal of the tracking focus IC but not at the output (TEO), the problem may be improper supply voltage to the IC, or a defective tracking-focus IC. Check the focus (FE) signal input and output before you conclude that the IC is the defective component.

To perform the tracking offset or balance adjustments, connect the scope to the TE signal test point and play a test disc. Adjust the SVR control so that the TE signal waveform is vertically symmetrical relative to OV. You usually won't need to perform the tracking gain adjustment unless you have replaced the focus optical assembly or control.

Adjust the tracking gain so that the waveform is not varying or jumping around (no large waves). Adjust the tracking gain adjustment only after the focus and tracking adjustments are made. To make this adjustment, connect the oscilloscope across the tracking coil terminals and adjust tracking coil gain control for 1.8Vpp to 2.2Vpp. Follow the manufacturer's tracking offset and gain adjustments.

Waveform 6 - Tracking Coil

When the CD player is first turned on, the tracking and focus assembly can be seen searching, even if the player shuts down. Often, this indicates that the focus and tracking assemblies are normal. The tracking offset (TEO) signal can be scoped at the TEO terminal of the focus-tracking-sled servo IC. The tracking signal is fed to a tracking driver IC or two driver transistors. The driver output signal is applied to the tracking coil winding. A waveform taken at the tracking coil indicates if signal is present (*Figure 22-10*).

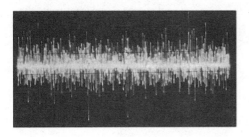

Figure 22-10. The tracking error (TE) waveform found across the tracking coil terminals.

Waveform 7 - No Motor Movement

Observation of the waveform at the motor terminals will reveal whether or not the drive signal is reaching the motor. If you observe a straight line, no signal is reaching the motor. Although a normal motor waveform will not show much movement on the scope, several lines may be seen together indicating that the motor is operating. A straight white line indicates no signal applied to the motor terminals (*Figure 22-11*).

Figure 22-11. No motor movement may show a white line waveform without any motor lines.

Waveform 8 - Sled Motor Movement

It is difficult to see a slide or sled motor operating while the disc is playing, as the motor is enclosed under the main PC board. A quick waveform taken at the sled motor terminals will indicate if the motor is operating. If the motor is operating, you'll see several thin lines on the scope (*Figure 22-12*). If you don't see a waveform here, trace the motor wires back to the main board. Often, the motor wires are soldered or plugged into the main PC board.

Figure 22-12. The sled motor movement may be seen as varying lines of a scope waveform taken at the slide motor terminals.

Check the motor waveform at the motor plug connections. Sharp pointed instrument prods pushed down inside the plug area will produce a good waveform. Clip the ground lead of the probe to common ground, since most motor windings are at ground potential.

The slide or sled motor is controlled by a signal from the same focus-tracking and sled servo IC. The slide motor control signal may be connected to the sled IC amplifier and to transistors or IC drivers. In some sled circuits, there may be a push-pull driver transistor in each leg of the sled motor terminals. In this case, one side of the slide motor terminals is not grounded.

Waveform 9 - Spindle Motor Movement

The disc or spindle motor rotates the disc platform. The spindle motor is controlled by a signal from the digital signal processor and CLV servo IC. The disc or turntable motor is a variable speed motor: it starts at around 500rpm when the laser pickup is at the center of the disc, and slows down as the laser pickup moves toward the outer rim of the CD (to approximately 200rpm). The master clock and spindle motor control circuits are fed to a spindle motor drive circuit that may contain two driver transistors or IC CLV servo. The disc motor is locked in with a correction signal from the CLV circuitry. Defects in a spindle motor may cause it to be dead and not rotate, or may cause it to operate at the incorrect speed. The waveform applied to the spindle motor of a Sanyo CD player is shown in *Figure 22-13*.

Figure 22-13. The spindle motor waveform may form lines close together and move up and down as the motor rotates.

Waveform 10 - Oscillator Waveforms

By observing the waveforms of crystal controlled oscillators waveforms in the digital servo and RAM circuits, you may be able to determine if supply voltage is applied to the IC processor and the component is operating. When the crystal-controlled oscillator is not functioning, you may assume the defect may be in the IC processor circuits.

The 4.19MHz crystal oscillator pins 31 and 32 of system control IC and 8.4672 MHz crystal oscillator on pins 53 and 54 of digital control processor in a Sanyo CP660 CD player may indicate IC302 and IC401 are normal (*Figure 22-14*). The timing control generator oscillator on pins 53 and 54 of an Alpine 5900 auto CD player may indicate a normal digital signal processor and CLV servo IC502.

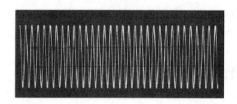

Figure 22-14. A 8.4672MHz crystal oscillator waveform found at pins 53 and 54 of a Sanyo CP660 player.

Conclusion

Observation of waveforms throughout the compact disc player may help the technician quickly isolate a defective circuit or component. The most important waveforms found in the CD players are the RF or EFM waveforms. Without this EFM waveform, the compact disc player will not operate.

Confirming that the laser diode is normal using a laser power meter test makes easy CD servicing. Check the manufacturer's service literature for correct laser offset and gain adjustments. The manufacturer may use special adaptors and alignment procedures, so check the service literature.

Chapter 23

Troubleshooting the CD Laser Head Assembly

By Homer Davidson

When a CD player is brought in for service, the condition of the laser head assembly may be the factor that determines if the unit is worth repairing. The CD laser assembly is so costly that, if it's defective, many owners will replace the compact disc player instead of having it serviced. The technician can save a lot of valuable time and trouble in CD servicing by first making sure that the CD head assembly and beam assembly are functioning, by performing a few quick laser head tests (*Figure 23-1*).

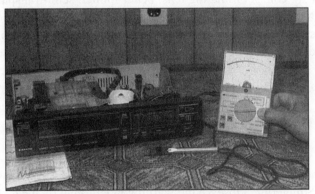

Figure 23-1. *The portable laser power meter will accurately indicate if the laser diode is emitting signal against the underside of the CD disc.*

Clean Up Time

Before attempting to measure the laser head beam, clean off the lens assembly. Dirt and dust on the lens assembly may produce skipping and poor tracking. In most CD players, the laser head assembly is covered by a flapper or disc assembly and cannot readily be seen. Of course, in the portable and boom box combination CD player, the lens assembly stares right back at you.

Clean the lens with a lint-free cotton swab or cleaning paper found in CD cleaning kits. Camera lens cleaning solution is ideal. Wipe the lens gently to avoid damaging the supporting spring. Blow the excess dust away from the optical lens with a can of compressed air dust spray, available at electronic and camera outlets.

Safety

The laser head assembly is a delicate and critical component that should be handled with care. Keep fingers and tools away from the optical lens assembly. Remember, theinfrared laser light is invisible. You can't see it. The laser assembly can be dangerous to your eyes. Never look directly into the laser beam or into the lens assembly. Keep your head at least 30cm (12 inches) from the lens assembly. Observe the warning labels found on the CD player.

The optical laser pickup assembly typically consists of the objective lens, focus-tracking coils, collimating lens, beam splitter, semi-transparent mirror, phot detectors, monitor and laser diodes. While working around or removing the laser assembly, wear a grounded wrist strap to prevent electrostatic discharge damage. Static electricity can damage the sensitive laser head assembly. The entire optical block is a single unit. When you encounter a defective optical assembly, you have to replace it as a unit.

Laser Head Properties

The laser head diode material may consist of GaAs/GAA1As, Ga-A1-As, or GA-AL-AS material. Observe the warning label found on the CD player. Blow the excess dust away from the optical lens. The wavelength of the laser head diode may vary from 750nm to 820nm. The laser power output may vary from 0.15nW to 0.7nW (*Figure 23-2*).

CD PLAYER	DIODE MATERIAL	WAVELENGTH	LASER OUTPUT
ALPINE AUTO CD 5900	GaALAs	780nm	0.4mW
ONKYO DX-200	GaAs/GaAlAs	780nm	0.4mW
PIONEER PD7010	Ga-AL-As	780nm	0.26mW
SANYO CP680		775nm to 830nm	0.7mW
SONY CDX-5	GaAlAs	780nm	0.4mW

Figure 23-2. The laser diode material, wavelength and laser power output of the various CD players.

There are several ways to test the laser head emission:

• an infrared card indicator,

• current and voltage measurements

• observation of the EFM waveform

• the laser power meter

The easiest method to detect whether the laser is emitting an infrared signal is with the infrared indicator card, but the card does not indicate the amount of radiation emitted. The current and voltage checks will give the correct current drawn by the laser diode, but it is necessary to get inside the unit to make this measurement. The laser power meter is the best way to make this measurement: the meter is easily attached, and you can see the actual radiation measurement on the meter display.

Infrared Card Indicator

The infrared indicator is a plastic card that was designed for testing infrared remote controls. The card contains a small square that glows when it is irradiated by infrared. Expose the infrared card to ordinary light for at least 5 minutes before use. The small sensitive square area must be held directly over the lens area for correct test results.

The infrared indicator card will indicate infrared emission, but will not indicate the intensity, or tell you if the emission is sufficient to operate the optical lens pickup assembly. This infrared indicator card can be purchased at most electronic distributors, Radio Shack and RCA. The RCA part number is 153093.

Current and Voltage Checks

The laser diode emission checks can be made by measuring the amount of current drawn by the laser diode. Most CD players have a current label attached to the assembly. For instance, the laser label may read 0138-499, indicating the current value is 49.9mA. Remember, the current increases when the temperature rises and decreases when it drops. If the current that you measure is much greater than the current specified on the label, the APC circuit may be defective, or the laser diode may have deteriorated. In either case, the laser head assembly must be replaced.

In some CD chassis, a jumper wire may offer a test point where a current reading can be obtained. Remove the plug and take a current measurement of the laser head diode. The current should match that shown in the CD service literature or on the laser power label.

Critical Voltage Tests

Another method to check the laser diode current is with a diode voltage measurement. Actually, in this case, the current drawn by the laser diode is determined by measuring the voltage across a fixed resistor in the laser drive emitter transistor circuit (*Figure 23-3*). Often, this voltage is less than 1V. Check the laser drive current by measuring the voltage across R209 (12Ω). Then compare the voltage measurement to the current label on the pickup assembly.

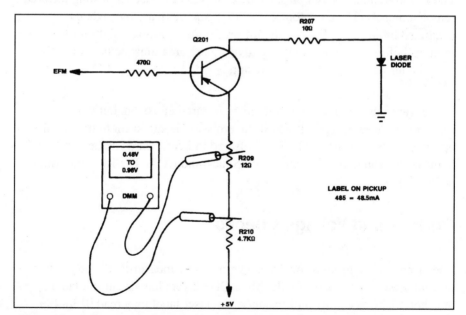

Figure 23-3. Voltage measurements across the emitter resistor can be converted to current, indicating if the laser diode is functioning.

The current drawn by the laser diode equals the voltage divided by the resistance. In this case, where the resistance is 12V, simply dividing by 10, or moving the decimal point one place to the left, is a close enough approximation. If the voltage across R209 is 0.48V, the current should be roughly .048 A or 48mA. The label on the pickup shows a reading of 485, which indicates 48.5mA.

In this case, the laser pickup head is normal. If the voltage measurement was IV, the current would be too high and might indicate a defective laser diode. When the current is 10 percent or more in excess of the recommended value, replace the pickup assembly.

EFM Waveforms

The eight-to-fourteen (EFM) modulation waveform must be present or the CD player will shut down after searching. The correct EFM or "eye pattern" waveform is found downstream from the rf amp IC or transistor (*Figure 23-4*). The presence of this waveform indicates that the laser head and rf circuits are normal. Most CD players have a test point at which the eye pattern can be observed with the oscilloscope. The EFM signal is fed to the signal and servo processors and IC.

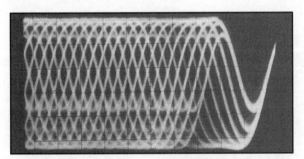

Figure 23-4. Presence of the EFM (eight-to-fourteen) waveform indicates that the laser optical assembly and rf amp are normal.

A low amplitude, jittery pattern, or no EFM waveform at all, may indicate a defective laser head assembly or rf amp circuits. Critical voltage and resistance measurements on the rf amp transistor or IC may determine if the rf amplifier is defective. When you find an improper EFM waveform at the rf amp but measure normal voltages on the rf transistor or IC, suspect a defective laser head assembly. In some cases you may have to replace the rf amp IC in order to determine if the EFM signal is present.

Although the presence of an EFM waveform indicates that there is an rf signal from the laser head assembly and the rf circuits, it does not indicate if the laser is drawing excessive current or has correct emission. The card indicator may indicate that infrared signal is present, but how much? Voltage and current measurements of the laser head circuits may provide a better indication of a defective laser diode, but this requires disassembly of the CD player. The quickest and best method to test the laser diode emission is with the laser power meter.

The Laser Power Meter

Most CD player manufacturers recommend a certain laser power meter or provide them through the various electronic depots. I use an LPM 5673 Sanwa laser power meter (*Figure 23-5*). This low-priced meter is available through distributors.

Figure 23-5. *The laser power meter may also be used to test infrared TV and VCR remote scope. The EFM signal is fed to the signal transmitters and servo processors and IC.*

This meter offers all required functions for the control of laser light sources. This instrument was particularly designed for service of the compact disc and laser disc players, with a narrow and tiltable probe. The instrument can be used on helium-neon lasers, such as used in distance meters and linear measuring instruments.

Besides CD player tests, the tester can check the function of cassette compartment LEDs in video recorders and the transmitting diode in infrared remote controls. The small black probe can be held against the end of the infrared remote to determine if the remote is emitting the infrared signals.

There are three measuring ranges: 0.3mW, 1mW and 3mW, which can check all infrared lighting sources found in consumer electronics. The meter display includes three measuring ranges of measured light output with the upper scale at 0.1mW, intermediate scale 0.3mW, and an extra scale for the display of the condition of the battery that powers the instrument.

A wavelength selector switch will set the value length of the laser light at 633nm, for helium-neon-red lasers, and at 750nm to 820nm for CD and LD players for infrared measurement. The function switch determines the desired function and measuring range. Also, the function switch turns the laser meter off and on.

Using a Laser Power Meter

To use this meter, connect the test probe to Jack 8. Set the function selector switch to the "Batt Test" position. If the meter hand stops in the good range, the batteries are normal and the tester is ready to check infrared light. The meter indicator will be deflected when the probe is turned toward sunlight or a service bench light. The probe should not be directly pointed at the sun or strong light, or the meter and probe might be damaged.

Check the battery condition each time the laser power meter is used. Observe the manufacturer's safety instructions in the service manual of the CD player. The specified laser output is listed in most CD manuals. Set the required measuring range with the function switch. Set the switch to the correct wavelength at 750nm to 820nm for CD or LD players.

In cases where the specified output of the laser diode in the CD player is not known, begin with the meter set to the highest range. Most CD player diode outputs are below the 1mW range. Rotate the switch to the 0 to .3mW range. Position the probe with the round window over the CD lens assembly. You may have to move the black probe around to get the maximum reading.

If the probe is positioned at an angle to the light beam, reflections will occur and result in a distorted measurement. Strong overhead light may cause an improper reading, so make your measurements away from the service center's main lighting. When the round hole of the probe is directly over the lens opening, the laser diode measurement will be maximum (*Figure 23-6*).

Figure 23-6. *In most CD players, the infrared protecting interlock system must be defeated before the laser diode comes on.*

Keep the window of the probe clean of lint, dirt and smudges. If the sensor window is soiled, you'll get a distorted measurement. Clean off the sensor window with a cloth or swab moistened with the same solution used to clean the lens area of the CD player.

Read the measured value on the laser power meter. Compare this measurement with that given in the service manual for the compact disc player. Adjust the laser power output according to the manufacturer's service manual. If the service manual is not available, compare the readings with similar compact disc players.

Defeated Interlock

CD players have an interlock system that prevents the laser from operating unless there's a disk in the tray and everything is properly closed up. This interlock system must be defeated during service in order to cause the laser diode to emit infrared light. Notice in *Figure 23-6*, in the case of this Sanyo CD player, a piece of cardboard is placed between the LED and the disc hole area keeping the LED light from striking the interlock sensor, thus providing operation of the CD player functions. Place either a piece of plastic tape or cardboard between hole and LED to make the player function. Remember, the laser beam is on when this interlock is defeated. Keep your eyes away from the lens area.

In most portable CD players, the top lid engages an interlock switch that turns on the player and laser diode when the lid is closed (*Figure 23-7*). A small screwdriver blade can be used to push down and engage the interlock switch, so that the intensity of the infrared light from the laser diode in this RCA CD portable may be measured.

Figure 23-7. The plastic lid of the portable CD player provides interlock protection as the lid must be closed before the unit will play.

Use a toothpick or a plastic pointed tool to defeat the interlock switch in other portable models. Clipping a wire across the interlock switch is another way to provide CD player action when the unit is opened up for service. Do not forget to remove the defeated interlock switch after repairs are made.

Testing Remote Control Units

The laser power meter may be used to check infrared remote control units as well. If the plastic pickup is placed against the infrared source, the 1mW or 3mW range should be used (*Figure 23-8*). Comparison measurements of the defeated infrared remotes can be marked or indicated on the meter scale. A weak or defective infrared remote transmitter may provide insufficient or no measurement on the laser power meter.

Figure 23-8. *Check the suspected remote control with the infrared laser meter using the higher power ranges.*

Chapter 24

Troubleshooting Audio Circuits By the Numbers
By Homer Davidson

Troubleshooting the sound circuits in a TV set is usually not difficult. The cause of dead audio circuits, as with dead TV chassis, is usually easily located. Troubleshooting weak and intermittent audio problems becomes more difficult. Often the cause of distorted sound is a malfunction in the audio output circuits.

Some technicians prefer to first check the voltage output of the power supply that supplies power to the audio stages. Others try to locate the dead sound stage by checking each transistor with in-circuit tests, while other technicians signal trace the audio using an oscilloscope and an external amplifier. If the audio is extremely weak, the cause of this symptom can be located by signal tracing using the scope as a monitor. Sound circuits can easily be serviced by the numbers.

The Early Audio Circuits

In early solid-state TV chassis, the audio circuits contained af, driver and single-ended output transistors. In later sets, the IC IF/Detector and preamp IC were introduced, then the audio output circuits of an af or driver transistor and two output transistors in push-pull operation. Transistor audio output circuits were quite popular at one time in most consumer electronic products. Later the audio output IC appeared.

IC components found in the audio stages consisted of not only the audio output circuits, but preamp and driver circuits. This same IC may also include part of the IF/SIF and discriminator sound circuits (*Figure 24-1*).

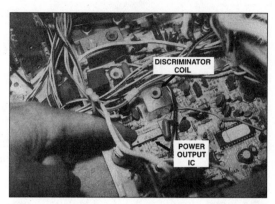

Figure 24-1. *The audio circuits in early solid-state TV sets consisted of a single integrated circuit mounted on a heat sink.*

Often the IC became leaky, causing dead, intermittent or distorted audio. The audio output terminal of the output IC was coupled directly to the permanent magnet (PM) speaker. Defective speaker coupling capacitors were known to produce weak, intermittent or distorted sound, or no sound at all. Simply replacing the audio output IC solved many different sound problems.

Today's Sound Circuits

Transistors are back in today's low priced TV chassis. You will find the same af amp and two small output transistors in push-pull operation. Usually, these transistors are found in the small portable or table-top model TV's.

Monophonic TV sound circuits can consist of one large IC, sound IF detector, and preamp in the large IC with additional sweep, AGC and color circuits, or a single power IC component. The sound output IC is mounted on a separate heat sink (*Figure 24-2*).

A stereo sound system may consist of a stereo demodulator circuit, demodulator IC, matrix, audio switching IC, and dual audio output IC. In other stereo chassis, the audio is capacitance coupled to a MPX/stereo IC, with left and right audio output terminals. The stereo signal is coupled to a Video/Audio control IC, dual-af amp IC, and to a dual-audio output IC.

Figure 24-2. All of the audio output circuits in some sets are included in one IC.

No Audio

Complete absence of sound output may be caused by almost any working component in the audio circuits. A defective relay, a blown fuse, a leaky or shorted output transistor or a leaky or shorted IC can result in the absence of sound. Open or leaky driver transistors can kill the sound, as can leaky or shorted IC can result in the absence of sound. Open or leaky driver transistors can kill the sound, as can leaky audio output IC or an open speaker electrolytic capacitor. Be sure to check bias resistors and diodes when trying to locate a dead transistor or integrated circuit.

Other malfunctions that can cause the absence of sound are, improper power supply voltages to the audio circuits or leaky electrolytic filter or decoupling capacitors. Check for open isolation resistors between the power source and the output transistor or IC components if there is no sound but voltage across the electrolytic filter capacitor is normal.

Weak sound is often caused by open or dried-up coupling capacitors. Especially, suspect electrolytic coupling capacitors of 1 mF to 10mF. Weak and garbled sound may be caused by misadjustment of the sound or discriminator coil. Open sound take-off coils can result in normal audio on local strong TV stations and weak sound on distance reception. Check the electrolytic speaker coupling capacitors for weak audio. A combination of weak and distorted sound can be caused by a leaky output transistor or IC. Audio output transistors and IC components are the most frequent cause of distorted sound, Sometimes leaky af or driver transistors or a leaky if/detector and preamp IC can be the cause of weak and distorted sound. Slight distortion with hum in one channel may be the result of a defective dual-output IC. Check all bias resistors and look for leaky capacitors tied to the audio output transistor or IC. If the audio sounds garbled, don't overlook the possibility that the if or discriminator coil may be misadjusted.

Intermittent Sound

Intermittent audio is the most difficult audio problem to locate and can take up a great deal of service time. A defective relay can cause intermittent sound or complete absence of sound. Soldered board connections and cold solder joints cause many intermittent sound problems. Intermittent audio can be caused by poor IC or transistor terminal connections to the printed circuit board. If you suspect that this is the problem, solder all terminal connections with fresh solder. Another possible cause of intermittent sound problems is intermittent transistors or integrated circuits.

Check By Numbers

The cause of most sound problems can be located quickly using routine checks. The first test is to make sure the volume control is turned up. Place your ear close to the speaker to determine if there are any signs of weak audio or audio hum. If hum is present, this indicates that ac voltage is possibly being applied to the audio output stage. Then, visually inspect audio circuits for burned or overheated components.

Check the audio stages by the number (*Figure 24-3*) to determine if the sound problem is within the audio output or in the IF/SIF, or discriminator circuits.

• Step 1: check for audio at the volume control by observing the audio waveform on the scope probe or by applying the audio from this point to the input of an external amplifier.

• Step 2: if these tests show that sound is present at the volume control, check the voltage source at the audio output transistors or IC.

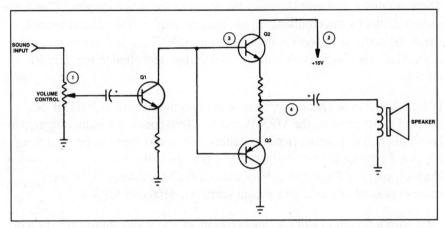

Figure 24-3. *An orderly procedure, checking "by the numbers" using signal tracing and voltage tests can quickly lead the technician to the cause of audio circuit problems.*

• Step 3: if source voltage is present at the audio output circuits, but there is no audio at the speaker, signal trace the audio at the base terminal of the output transistor or the input terminal of the IC.

• Step 4: check for the presence of signal output at the collector of the transistor or IC output terminal.

If no sound is found at the volume control in Step 1, proceed to the sound front end of the IF/SIF detector and preamp IC circuits for the following tests.

• Step 1: check the audio output at the output pin terminal of the SIF IC with a scope or external amp.

• Step 2: take a critical voltage measurement at the SIF IC supply terminal.

• Step 3: if the picture is normal, check the sound IF/SIF input terminal of IC with the scope demodulator probe.

Audio Signal Tracing

One good way to isolate the cause of audio problems in a TV set is to signal trace the audio circuits with a TV station tuned in using an oscilloscope or an external amplifier as the audio monitor.

There are many different stages to check out in the stereo circuits. Check the audio circuits by the numbers. If you find no audio at the volume control, check the audio waveform at the output pin of VIF/SIF IC. When no sound is found at this point, check voltages and components tied to the VIF/SIF circuits.

If the waveform or sound is present at the output terminal of VIF/SIF IC, proceed to the input of the MPX/stereo IC. Now check the audio output at both right and left output stereo channels. The audio amp can be used from this point on with the external amplifier. If there is no sound on either output channel, check voltage and components on the MPX/stereo IC. If one channel is dead or weak, you should suspect a defective MPX/IC.

Next, check the input audio at the pre-amp, af and audio control IC, if there is one in the circuit. Both channels of audio can be signal traced with an audio signal from stereo/MPX IC, through control IC, preamp, af and volume control circuits. Signal trace with a scope and an external amp. Likewise check the audio output stages by the numbers.

No Sound—Normal Color Picture

An RCA CTC146E chassis had no sound output. Step 1 of the audio checking procedure revealed that there was audio at the volume control. This suggested that the trouble was in the sound output circuits.

I measured the voltage as recommended by Step 2. The voltage at the collector terminals of the output transistors was zero (*Figure 24-4*). Absence of the 18V audio circuit source indicated problems in the low voltage power supply.

The schematic showed that the 18.5V source is taken from a low voltage derived from the flyback. A quick voltage measurement at the cathode terminal of the silicon diode CR4120 and filter capacitor C4135 (47μF) indicated normal voltage. Further checks in this area revealed that resistor R1211 (5.6W) between the 18.5V source and the collector terminal of Q1202 was open. Both the output transistor and R1211 were replaced to insure proper audio to the 32W speaker. This sound problem was easily located with step 2 of the checking procedure.

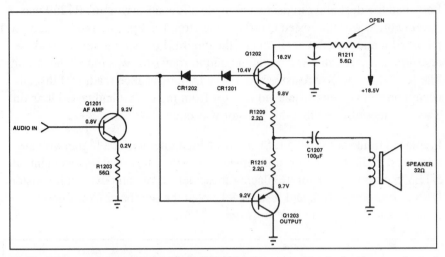

Figure 24-4. In one RCA CTC146E set, R1211 was open, resulting in the absence of sound.

No Right Channel Audio

In an Emerson M250 set, the left channel was normal but there was no audio in the right channel. I checked the audio at the volume control of the right channel (Step 1). There was normal sound at the volume control, so I checked the voltage at the collector terminal of the output transistor (Step 2). The collector terminal of transistor Q1206 was at ground potential. The collector terminal of transistor Q1205 should be at around 100V, but it measured only 47.7V (*Figure 24-5*).

Figure 24-5. Careful voltage measurements at the pins of integrated circuits or transistors can help locate the defective part.

I brushed the dust and dirt aside and made a few observations. R1259 was running quite warm. I tested Q1205 in the circuit. It appeared to be leaky, so I removed it and tested it again out of the circuit. I decided to test all bias resistors that were checked while the output transistor was out of the circuit. The R1260, an 8.2KΩ resistor had a very high resistance reading. I disconnected one end of this suspected resistor from the circuit and tested it again. This check confirmed that this resistor was open.

I replaced transistor Q1205 (2SC3296), which operates at a higher voltage than most audio transistors, and replaced it with a universal replacement; an NTE375 output transistor. Replacing these defective components; transistor Q1205, resistor R1260, and resistor R1259 at the number 2 check point restored sound output in this set (*Figure 24-6*).

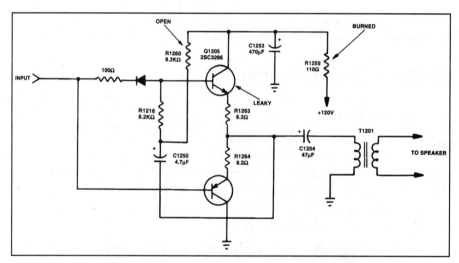

Figure 24-6. *In an Emerson MS250 TV set, replacement of leaky Q1205, open R1260 and R1259 restored the sound.*

Weak and Distorted Right Channel

An RCA CTC 166 had a weak and distorted sound in the right channel. When servicing audio problems in stereo audio systems, you will generally find that it will take a little longer to locate the defective component, since there are many different circuits to check out.

Since only the right channel in this set was weak and distorted, I skipped diagnostic step number 1 and went directly to the audio output circuits. A check of the schematic diagram for this set revealed that this RCA CTC 166 chassis has a volume control IC (U1801) feeding a dual audio output IC1900.

I measured the voltage at pins 1, 12 and 13. Naturally, the supply voltage at pin 12 and left output channel 13 should be normal since the left channel audio was normal (Step 2). The voltage at pin 1 was only 7.2V, indicating a possible defective output IC (*Figure 24-7*).

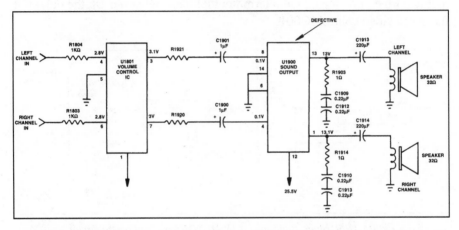

Figure 24-7. Integrated circuit U1900, which had a leaky right channel, was replaced with an exact replacement part to solve an audio problem in this RCA CTC166 chassis.

I applied the input audio from pin 4 to an external amplifier, and compared it with the audio at the normal left channel at pin 8. I found that the right channel signal produced a lower volume in the external amplifier. To make sure that the audio signal was present ahead of the output IC, U1900, I tested the right channel audio at pin 7 of U1801. Both audio channels were fairly comparable at pins 3 and 7 of U1801. This led me to believe that the right channel circuitry of U1900 was defective. I tested all components that were connected to each pin of U1900. They all appeared normal. I ordered a replacement for U1900, an RCA 181836 exact replacement part. When this replacement was installed, it corrected the weak and distorted channel.

Intermittent Sound-Left Stereo Channel

Intermittent sound problems in stereo channels can consume a great deal of service bench time. For example, in the Goldstar CMT-2612 that I was working on, an intermittent left audio channel can be caused by problems at any point from the MPX/Stereo IC601, through the Audio/Video control IC001, preamp and AF IC901, to audio output IC902 (*Figure 24-8*). Since there were so many different audio stages, I decided to check the circuits one half at a time, monitoring the intermittent left channel at pin 18; the output of the audio video control IC001.

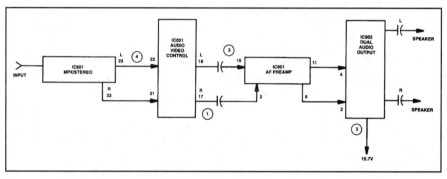

Figure 24-8. Block diagram of a TV MPX/Stereo sound output circuit.

After about an hour of operation, the left channel speaker became a little noisy. Pin 18 of IC001 showed no signs of intermittent audio. This meant that the intermittent component must be from pin 18 towards the speaker. Since a known-good test speaker was used in testing, the problem could not be the speaker. It had to be an intermittent component in the left channel circuitry.

I divided the remaining suspect stereo audio circuits in half. By checking with the test monitor at pin 11 of IC901, I found that the left channel was still intermittent at pin 11, indicating that the problem was between pin 18 of IC001 and output pin 11 of IC901. I connected the scope probe to pin 15 of the left channel at IC901 (*Figure 24-9*). The left channel can also be monitored at the left channel output jack that connects to the input circuits.

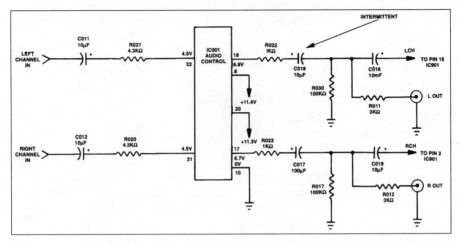

Figure 24-9. *Replacement of an intermittent capacitor, C016 (10µF), in the left channel in a Goldstar CMT-2612 TV corrected an intermittent sound problem.*

After several minutes, the left channel sound began to act up. The waveform at pin 15 of IC901 reflected the problem. This meant that the intermittent part was either R022, C016 or C018 from pin 18 of IC001 to pin 15 of IC901. When I sprayed coolant on C016 (10mF), an electrolytic coupling capacitor, the sound cleared up. Capacitor C018 was found to be intermittent. Replacement of C018 solved the problem.

Conclusion

Sound problems can be serviced by the numbers starting at the volume control. Some technicians may prefer to start at the SIF/Detector IC and work towards the speaker. I prefer to cut the audio circuits in half and start at the volume control to determine if sound problems are in the input or output sound stages. You will find that when the problem is weak or intermittent sound, or if the audio is stereo, it will take a little longer to locate the defective component.

Chapter 25

Compact Disc Interactive (CD-I)—Part 1

By Marcel R. Rialland

CDI, or compact disc interactive can be described as a multimedia system that is capable of delivering audio, graphics, pictures and text interactively. The term "interactive" means that instead of simply listening to music or watching a movie, the user can interact with the system to alter the order in which the system retrieves the disc information, and which portions of the information will be retrieved. The Philips CD1910, shown in *Figure 25-1*, was introduced last October as the first consumer CDI player. This player can also play standard digital audio compact discs (CD-DA) as well as the new Photo CDs.

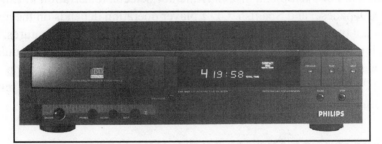

Figure 25-1. *CDI, or compact disc interactive is a multimedia system that is capable of delivering audio, graphics, pictures and text interactively from a compact disc. This CDI player can also play digital audio CDs as well as photo CDs.*

The User Interacts With the System

The CD-I player basically uses the compact disc format as a storage medium for both audible and visual information, as well as text and control data. More importantly, it provides interactivity for the user. For example, the user may use a CDI system to learn to play a musical instrument, learn a language, "visit" a museum, or play an interactive game. Software in the area of education (interactive training), entertainment, information, and reference are available in the consumer market. In addition, the CD-Interactive system has had commercial applications in business and industry. Because of the software demands of these applications, expanded forms of formatting information on the disc had to be developed. It also means CDI players require additional decoding circuits.

The CD-I Operating System

The CD-I operating system is the compact disc Real Time Operating System (CD-RTOS), based on the OS-9 operating system. CD-I software enables synchronization of audio and video information through the interleaving of digital audio and video data on the disc. CD-I may combine audio, video (stills or moving), and text in a single application. For example, a CD-I application may consist of a narration (audio) along with text on the screen while a picture (video) is displayed on the monitor (standard TV monitor). Another application may use animation in sync with the audio. CD-I also allows for the selection of one of several languages, depending on the application. For example, a disc may include selectable narration in English, French, Japanese, and Spanish. CD-I player operation depends on the application and type of disc. All compact discs have some common features, including error correction, interleaving, EFM (Eight-to-Fourteen Modulation), and a storage capacity of up to 650MBytes of digital information. *Figure 25-2* illustrates the compatibility of each disc type. Let's now review and compare each disc type.

Compact Disc-Digital Audio (CD-DA)

CD-DA is the most familiar and popular application of digital compact disc. The compact disc is recorded to provide high fidelity audio with virtually no distortion or noise. The CD-DA format is the basis for all other CD formats. CD-DA makes use of 16-bit PCM (Pulse Code Modulation) to place data on disc. In the encoding process, the analog audio is converted to 16 bits per channel at a sampling rate of 44. 1 kHz. Each 16 bit sample is then divided down to an eight-bit symbol.

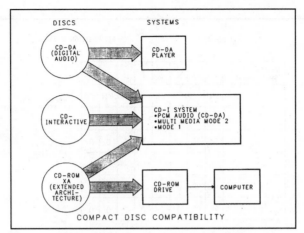

***Figure 25-2.** Compact disc compatibility.*

The CD-DA encoding process arranges six stereo sample periods of 192 bits or 24 bytes (6 samples times 32 bits, or 4 bytes for right and left audio) into a frame, known as a Small Frame. A Control and Display code (subcode data), parity codes, merging bits, EFM, and a sync code are all applied to the six stereo samples. Thus, a CD-DA small frame consists of 588 bits. This results in a frame frequency of 7.35kHz and a bit clock frequency of 4.3218 megabits per second.

Ninety-eight small frames make up a Large Frame or Subcode Frame (see *Figure 25-3*). The subcode repetition rate is 75Hz. The Subcode Frame is equivalent to a CD-ROM sector, which contains 2352 Bytes of data (98 small frames times 24 Bytes). The subcode is necessary to provide the CD player with information such as elapsed time and control data as illustrated in *Figure 25-4*. There are eight channels used in the Frame format, labeled P through W. The lead-in track contains the Table Of Contents (TOC) information, incorporated in the Q-channel. The CD-DA format specifications limit the total playing time to 72 minutes of high-fidelity stereo.

Compact Disc Read Only Memory (CD-ROM)

CD-ROM is another type of disc based on compact disc technology. A CD-ROM disc may contain more than 600 megabytes of data. CD-ROM defines data in the form of sectors. Each sector contains 2352 bytes of information and is recorded using the same EFM (Eight-to-Fourteen Modulation) technique used in CD-DA. EFM provides a first level of error protection well suited to audio data as well as binary data in general.

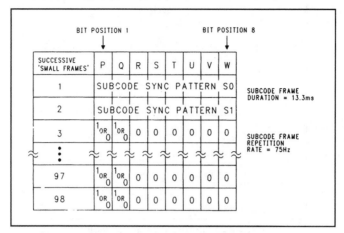

Figure 25-3. Subcode frame format.

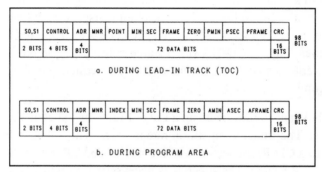

Figure 25-4. Q-channel format.

The sector contains synchronization, address and mode information. In addition, a sector contains a user data area of either 2048 bytes for Mode 1 or 2336 bytes for Mode 2 (See *Figure 25-5*). The difference between these two modes is that Mode 1 uses 288 bytes to provide an additional level of error detection (ED) and error correction (EC). This ensures a level of data integrity essential for critical information that does not degrade gracefully, such as text and binary data typically contained in databases.

Mode 2 trades this benefit of additional data security for a maximum data transfer rate by making the additional 288 byte area available as user data. In this case EFM is adequate for error protection of data such as video and audio. The standard CD Table of Contents (TOC), although not available to the computer program, may be used by the CD-ROM player to locate a requested track. The TOC appears in the Q channel in the lead-in area of each disc. There are two types of tracks that the TOC can identify as stored on a CD-ROM disc: CD digital audio tracks, and data tracks.

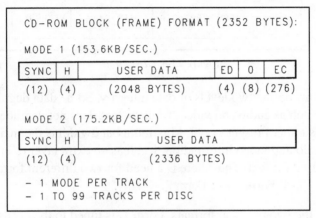

Figure 25-5. CD-Rom sector.

Compact Disc Interactive (CD-I)

CD-I specifically meets the needs and requirements of the CD-I player. Since CD-I information may include audio, video, text, and program data, several different encoding methods are used. Encoding standards are established for CD-DA, CD-ROM, and CD-I. Formats may be mixed on the disc, but Track One must always identify the disc as CD-I.

As with CD-ROM, CD-I defines data in the form of sectors. Each sector contains 23 52 bytes (see *Figure 25-6*). The CD-I physical format is based on CD-ROM, Mode 2. CD-I is primarily an audio/video driven medium. Thus, video must be synchronized to the audio with the CD data rate of 75 sectors per second. At the same time, there is a need for real-time interactivity. Thus all three data types, audio, video, and text (binary data), are physically interleaved. The sub-header (SH) mechanism is used for real time physical interleaving of data.

```
                          BASED  ON  MODE  2
        FORM  1:

        SYNC  HEADER  SUBHEADER   USER DATA   EDC   ECC
        12B     4B        8B        2048B      4B   276B


        FORM  2:

        SYNC  HEADER  SUBHEADER   USER  DATA    RESERVED
        12B     4B        8B         2324B        4B
```

Figure 25-6. CD-I forms.

The two forms define two levels of data integrity. Some data degrades gracefully, such as audio and video, whereas text does not degrade gracefully. Text is either present or not. Maximum bandwidth is the main requirement for audio and video information, whereas an extra layer of error correction is required for text. Thus there is a need for two different formats in Mode 2 for CD-I: Form 1 and Form 2.

The first of the two physical formats, Form 1, is tuned to the needs of text, computer data and highly compressed visual data. Thus, Form 1 uses 280 bytes for additional error detection and correction (Error Detection Code or EDC and Error Correction Code or ECC), leaving 2048 Bytes as user data. The second physical format, Form 2, is used to fill the requirements of real time audio and visual data, leaving 2324 bytes of user data plus 4 bytes of reserved data.

The CD player is designed so that the rotational speed of the disc can be varied to ensure constant linear velocity at the readout head, resulting in a constant data transfer rate (frame rate) of 75 sectors per second. The resulting data transfer rates are 153.6KBytes/s for Form 1 and 174.6KBytes/s for Form 2. Let's now look at what type of data can be encoded in the CDI format.

Audio Formats

The audio formats are illustrated in *Figure 25-7*. There are four audio formats that may be applied to CD-I. The first is the familiar CD-DA. The standardized format for encoding CD-DA as Pulse Code Modulation (PCM) includes the 16-bit samples (Significance), at a sampling rate (f_s) of 44.1 kHz. This results in a dynamic range of greater than 90dB with a bandwidth (BW) of 20kHz and a maximum playing time of 72 minutes of hi-fi stereo audio. This format limits the quantity of information which can be placed on the disc. Thus the CDI standard allows for three other audio formats.

FORMAT	fs IN kHz	SIGNI-FICANCE bits per sample	BW IN kHZ	CHANNELS	XSYS IN MHz	t MAX IN MIN STEREO (MONO)
CD-DA (PCM)	44.1	16	20	1 STEREO	11.2896	72
LEVEL A (AD-PCM)	37.8	8	17	2 STEREO 4 MONO	9.6768	144 (288)
LEVEL B (AD-PCM)	37.8	4	17	4 STEREO 8 MONO	9.6768	288 (576)
LEVEL C (AD-PCM)	18.9	4	8.5	8 STEREO 16 MONO	4.8384	576 (1152)

Figure 25-7. CD-I formats.

The audio data coding used in CD-I is 8 or 4-bit Adaptive Delta Pulse Code Modulation (ADPCM). A lower sampling rate and a different coding technique is used since no more than 50% of the time is allocated for audio information. The Adaptive Delta PCM (ADPCM) coding technique used to store audio information more efficiently, requires additional processing beyond 16-bit PCM for both encoding and decoding.

The chart of *Figure 25-8* shows the specifications for each level. The level used depends on the application. For example, to provide maximum time where high fidelity is unnecessary, such as a narration, Level C is used. Using this level limits the frequency response to 8.5kHz, but allows up to approximately 19 hours (with no other data: video, CD-DA, text) of mono audio or 9 + hours stereo on a single disc.

By using the three levels of ADPCM, information other than audio (video, text, and program) can be included on a disc, while still allowing 72 minutes of audio, as illustrated in *Figure 25-8*. The CD Information Intensity Chart shows the percentage of data which can be allotted for non-audio data for each level compared to CD-DA. Thus, 100% of a CD-DA disc is used when 72 minutes of audio is encoded onto the disc. If the same 72 minutes is encoded using ADPCM Level A, only 50% of the disc is used for audio, leaving 50% for non-audio data. Likewise, Level B allows 25% for 72 minutes of audio and 75% for non-audio data. Level C allows 60% for audio and 94% for non-audio data.

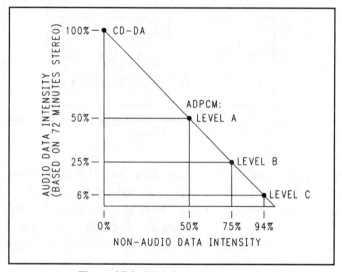

Figure 25-8. CD Information intensity.

Video Formats

Because there are several different television systems used around the world (see *Figure 25-9*), the video encoding system for CD-I allows for a world-wide standard. That is, the video data can be decoded to play on NTSC, PAL, or SECAM television systems. Besides the various audio quality levels, there is a need for various video quality levels. The video quality levels offer a choice of resolution and picture type.

```
+---------------------------------------------+
|              CD-I VIDEO FORMATS:            |
|                                             |
|   •  WORLD-WIDE FORMAT                      |
|      - NTSC, PAL, AND SECAM                 |
|   •  THREE RESOLUTION MODES                 |
|   •  THREE PICTURE TYPES                    |
|      (ENCODING METHODS)                     |
+---------------------------------------------+
```

Figure 25-9. CD-I video formats.

The resolution modes provide for both present and future television systems as illustrated in *Figure 25-10.* The three modes are Normal, Double or Enhanced, and High resolution. The chart shows the three modes with their respective horizontal and vertical lines of resolution for all three television standards.

	NTSC 525 Lines	PAL/SECAM 625 Lines
Normal	360X240 Pixels	384X280 Pixels
Double	720X240 Pixels	768X280 Pixels
High	720X480 Pixels	768X560 Pixels

Figure 25-10. Video resolution modes.

The picture code depends on the type of picture to be displayed. *Figure 25-11* compares each type of coding system. Picture coding provides for two picture quality levels: natural pictures and graphics. Natural stills are best handled by YUV (Y, R-Y, B-Y) coding for an equivalent of 24-bit color depth. Color Look-up Tables (CLUT's) provide high quality complex graphics. Absolute RGB coding is best used for user manipulated graphics. Run Length Encoding is used for text, graphic animations, and graphic images which require few colors in large areas of the screen. Compression techniques are required to provide full screen animation in the graphic modes.

FORMAT	APPLICATION	MEMORY	COLORS
DYUV	NATURAL STILLS	108KB/PICTURE	ALL
CLUT	GRAPHICS ANIMATION	108KB/PICTURE	256 of 16 Million
RGB	USER MANIPU-LATED GRAPHICS	215KB/PICTURE	32,768
RLE	GRAPHICS	10-20KB PICTURE	128

Figure 25-11. Picture types (Encoding Process).

Natural pictures, using YUV (Y, R-Y, and B-Y) coding, occupy about 325kB per picture without interlacing (650kB with interlacing). To decrease throughput times and maintain a high quality image, all natural pictures are compressed with DYUV (Delta-YUV) coding. DYUV reduces the memory requirements to 108kB/picture. Thus, the DYUV coding system provides a transfer rate of one full-frame in about 0.6 seconds at a data rate of 174.6kB/s (Form 2).

The CLUT (Color Look-Up Table) mode is used for graphics animation. CLUT can be used as 256 colors out of 16 million, requiring 108kB of storage capacity per picture. Compression can reduce this to less than 10kB per picture. CLUT with compression provides full-screen animation with the interleaving of pictures and sound. A picture refresh rate of 17 frames per second is achievable in Form 2.

The other graphics mode is based on absolute RGB coding and is applied to user manipulated graphics. Fifteen-bit RGB graphics (32,768 colors) produce exceptionally crisp pictures at a cost of about 215kB per picture. No compression is used in this encoding system.

Text Coding

Text encoding may by handled using two basic methods, by a bit map process or with character encoding as illustrated in *Figure 25-12*. The bit-map process requires five bytes for each character. This limits the number of characters to a maximum of 120 million per disc, if only 16 colors are used in an 8 x 10 matrix of any shape.

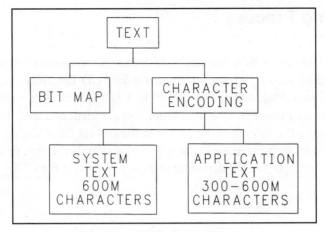

Figure 25-12. *Text coding.*

Character encoded text can be system text or application text. The standard character encoded text (system text), using one byte per character, allows 600 million characters in a full disc. Application text is encoded with two bytes per character. The second byte specifies factors like color, font type, and size. This extended coding method allows 300 to 600 million characters per disc.

There is a need to limit the number of characters on screen due to the limited resolution of a normal TV. Thus, text is limited to 40 characters on 20 lines. The characters are contained in a safety area of 320 x 210 pixels in the center of the screen. With the high-resolution screens used in computer monitors and future high definition or digital TV's, the High Resolution mode allows 80 characters to be presented on up to 40 lines. The safety area for the High Resolution mode is 640 x 420 pixels. The text is only stored once since compatibility between the two resolution modes is maintained.

Video Effects

A wide range of visual effects are provided in the CD-I system, including: wipes, cuts, scrolls, overlays, dissolves and fades. Up to five overlaying video planes are provided, with both transparency and translucency for all except the background plane. One plane is reserved for the background and another for the cursor.

Decoding Process

The CDI player must have the ability to decode information stamped on the disc. Decoding is straightforward in the standard CD-DA since it uses only one type of encoding method. However, the CD-I system uses more than one type of encoding process, which includes audio, video, and text. The data, once read from the disc, must be routed to the correct decoding circuits to be converted to its respective analog signal, whether audio, video, or text. Part 2 of this series will cover the decoding and control system for CDI.

Chapter 26

Compact Disc Interactive—Part 2

By Marcel R. Rialland

In the first part of this series of three articles, which appeared in the August issue, we described compact disc interactive (CDI), as a multimedia system that is capable of delivering audio, graphics, pictures and text interactively. The term "interactive" means that instead of simply listening to music or watching a movie, the user can interact with the system to alter the order in which the system retrieves the disc information, and select which portions of the information will be retrieved.

Compact Disc Interactive (CDI) software contains data in the form of audio, video, text, and control data. The first article looked at the applications of CDI and the method of formatting the different types of data. This article will examine how that information is retrieved and processed by the CDI player.

Microprocessor and Operating System

The microprocessor and operating system form the heart of the CDI control system, allowing real time operation and interactivity as required by CDI applications. The operating system also allows for control of CD-DA (Compact Disc Digital Audio) and Photo-CD. Thus the control system must determine the type of disc that has been loaded.

Real time applications require machine language to execute specific tasks. All machine language sets are specific to a microprocessor family. Specifying the microprocessor family and operating system makes it possible to produce discs carrying audio, video, text, binary data and application programs that will work on all CD-I players from all manufacturers. The microprocessor family specified for CD-I is based on the 68000 family. The Philips CDI910, CD1601 and CD1602 use the SCC68070 microprocessor.

The Compact Disc Real Time Operating System (CD-RTOS) used in CD-I is based on the OS-9 real time operating system. CD-RTOS is customized to fit the needs of the CD-I system. *Figure 26-1* illustrates the CD-RTOS organization.

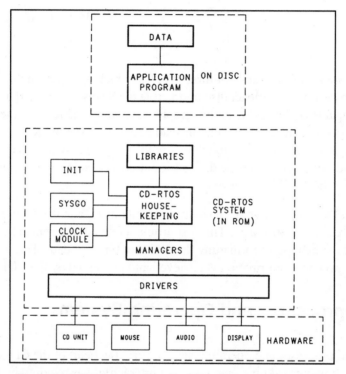

Figure 26-1. *The Compact Disc Real Time Operating System (CD-RTOS) used in CD-I, based on the OS-9 real time operating system, is customized to fit the needs of the CD-I system. CD-RTOS is organized as shown here. A series of instructions (CD-RTOS) is loaded from ROM into memory(booted)to create the user shell and load the operating system libraries, managers and drivers when the player is turned on.*

A series of instructions (CD-RTOS) is loaded from ROM into memory (booted) to create the user shell and load the operating system libraries, managers and drivers when the player is turned on. The user shell along with the peripheral devices, such as the mouse or the remote control, allows the user to interface with the system hardware and software.

User Shells

A Start-up Screen (*Figure 26-2*) is displayed when the player is turned On without a disc. The infrared remote control or other pointing device, such as a mouse, is used to make screen selections. A selection is made by placing the arrow cursor over a command icon and pressing the Activate Key (one of the keys around the joystick). Alternatively, dedicated keys, such as the Play or Open Keys, may be used to perform player functions. Other screens are accessed from the Start-up Screen, including the Info Screen (Help), Memory Screen and Settings Screen.

Figure 26-2. When the player is turned on without a disc in it, this Startup screen is displayed. A function selection is made by using the infrared control, mouse or other pointing device to place the arrow over a command icon and pressing the Activate key.

The Settings Screen (*Figure 26-3*) allows a user to set the date and time, as well as selecting options regarding the playing of discs. For example, the user may choose to turn the Auto Shuffle option On. Then, the Auto Shuffle option is activated any time a CD-DA is loaded into the player.

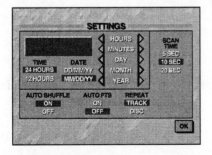

Figure 26-3. This Settings Screen allows the user to set the date and time, or select options regarding the playing of discs.

Figure 26-4 illustrates the user shell that comes up when a CD-DA (Compact Disc - Digital Audio) disc is loaded. All functions normally available on CD players are found in this shell. Tracks may be programmed for play or the whole disc may be played. Also, a programmed play sequence may be stored in FTS (Favorite Track Selection) for later playback.

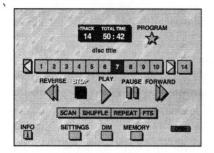

Figure 26-4. The user shell shown here is displayed when a CD-DA disc is loaded. All functions normally available on CD players are shown in this shell.

FTS programming also allows the user to enter a title. From then on, when that disc is played the title will appear above the Track Bar. The FTS may also be changed at any time by the user or may even be deleted. Selecting the Memory Icon allows the user to select and delete a disc from FTS.

The CD-I start-up screen (*Figure 26-5*) is displayed when a CD-I is loaded in the player. Clicking on "Play CD-I" begins the CD-I application. The next displayed screen, normally an introduction screen, is dependent on the software.

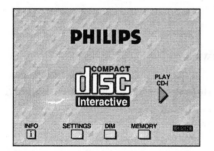

Figure 26-5. When a CD-I disc is loaded in the player, this CD-I Startup Screen is displayed. Clicking on "Play CD-I" begins the CD-I application.

CDI910 Processing

The major processing and decoding circuits in the CDI910 are contained on four circuit boards (see *Figure 26-6*):

• the CD Unit

• the MMC (Multi Media Controller) Unit

• the APU (Audio Processing Unit)

• the Video Encoder Unit.

Other assemblies found in the CDI-910 include:

• the Power Supply Unit (switch mode power supply)

• the Front Panel Assembly

• the RF Modulator/Switch

• the CDM-9 single beam compact disc mechanism.

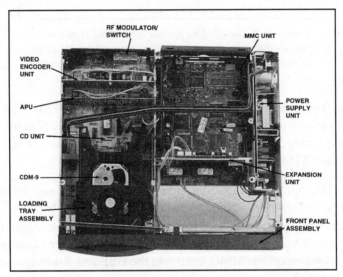

Figure 26-6. The major processing and decoding circuits in the CDI1910 are contained on four circuit boards, as shown here.

Let's look at the function of the four processing and decoding circuits.

The CD Unit

Before the information on the disc can be decoded, it must be read as in any compact disc system. So the CDI player is first basically a CD player. The CD Unit (see *Figure 26-7*) must optically read the disc, demodulate the EFM, regenerate the bit clock, control the speed of the turntable motor, and decode the interleaved data.

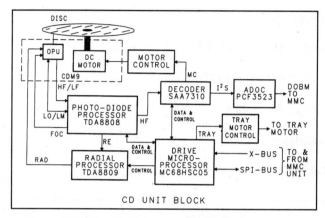

Figure 26-7. *The CD Unit portion of the CD-I system is the portion that optically reads the disc, demodulates the EFM, regenerates the bit clock, controls the speed of the turntable motor and decodes the interleaved data.*

The CDI910 incorporates the new CDM9 assembly to optically read the disc. The CDM9 is a swing-arm disc reading mechanism. This system uses single-beam disc tracking and operates in conjunction with the servo IC's, TDA8808 and TDA8809.

The Drive Microprocessor controls and monitors the servo and decoding circuits. The Drive Microprocessor also controls the player's tray motor. The microprocessor receives disc-access commands (for example: Jump to an absolute time, Pause, Stop, and Read-TOC) from the Master Microprocessor, MC68070 (located on the MMC Unit), via the DSP and X-bus (control and communications buses).

The Drive Microprocessor initiates the start-up sequence of the CD Unit. It activates the focus start-up circuit of the Photodiode Processor, controls the position of the CDM9 Swing Arm via the Radial Processor, and starts the CDM Motor by way of the Decoder SAA7310.

The Drive Microprocessor also monitors error conditions from the Photo-diode Processor and Decoder circuits. At the same time, the Drive Micropro-cessor sends messages (such as radial error or no disc detected) to the MMC Unit via the SPI-bus.

When the CD Unit is in the play mode, the OPU (Optical Pickup Unit) picks up the Low Frequency Signal developed from the wobble signal (generated in the Radial Processor) to make focus and radial corrections as the disc plays. Also, the HF signal (digital data) is picked up by the OPU and is amplified by the Photodiode Processor. The HF is coupled to the Decoder (SAA-73 1 0) for further processing.

Decoder Section

Decoding of the HF is accomplished by the Decoding Section (see *Figure 26-8*), which comprises four major active components:

• the Drive Processor (CD Drive Microprocessor)

• the SAA7310 Decoder IC

• the DRAM (MN4269-15) and

• the ADOC (PCF3523, Audio Digital Output Circuit).

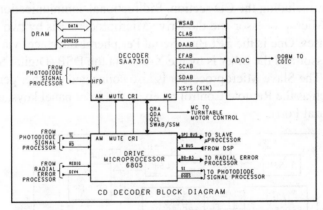

Figure 26-8. The Decoding Section comprises four major active components: the Drive Processor, the Decoder IC, the DRAM and the ADOC.

The SAA7310 Decoder IC incorporates the functions of demodulator, subcode processor, motor speed control, error correction, and error conceal-ment. The decoder accepts data from the disc and outputs serial data via the Inter-IC signal bus (IIS or I2S bus) directly to the ADOC IC. The I2S bus consists of three lines: WSAB (Word Select), CLAB (Clock), and DAAB (Data).

The Decoder IC also sends error codes via the EFAB (error Flag) line. In the case of CD-I, the error concealment function of the Decoder IC is disabled. This is because of the added error correction encoding included in the CD-I format. In addition to the I2S signal, subcode data is sent to the ADOC via the SCAB (Subcode Data) lines. The System Clock (XSYS) is also transferred to the ADOC IC.

The ADOC combines the I2S signal with the subcode data and the error flags and converts this signal into DOBM (Digital Output Bi-Phase Mark code) signal format. The DOBM signal from the ADOC includes not just the audio digital data, but also all the digital data (control codes, video data,text data, and program data) picked up from the disc. The DOBM data stream is applied to the MMC Unit for decoding.

MMC (Multi Media Controller)

The MMC Module is the heart of the CDI system (see *Figure 26-9*). The Master Microprocessor (68070) controls and manages all the activity in the CD-I player, including the CD section. Bidirectional communication between the Master Microprocessor and the Drive Microprocessor is by two communications buses. One is the SPI Bus (Serial Peripheral Interface) via the Slave Microprocessor and the other is by the X-bus via the DSP (Digital Signal Processor). The Slave Microprocessor is also connected to other peripheral devices, such as the Remote Control receiver, the front panel keys, and Port I (RS-232 Serial Port).

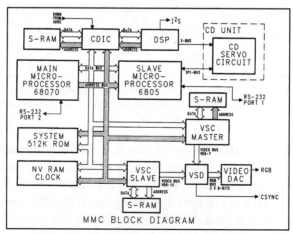

Figure 26-9. The MMC module is the heart of the CDI system.

When the player is first turned On, all of the microprocessors are reset. Next, the instructions from ROM are executed by the Master Microprocessor, setting up the operating system for CD-I. The start-up procedure for detecting and reading a disc is also followed. When a disc is detected, the TOC is read to determine the type of disc installed. The appropriate user shell is then displayed on the monitor. The front panel also displays the type of disc detected. The key MMC components and their functions are summarized below:

• CDIC (CD Interface Circuit): sends commands to the CD unit via the DSP and decodes DOBM (Digital Output) from the ADOC.

• DSP (Digital Signal Processor): interface between CDIC data and address lines and X-bus to communicate with CD unit.

• VSC's (Video and System Controllers): build images for planes a and b respectively; control access to video RAM and EPROM.

• VSD (Video Synthesizer): combines or selects inputs from both VSC'S. Also adds special effects (wipes, fades, dissolves, etc.) when switching between planes. Note: The CD1601/ 602 uses a VSR IC, which incorporates the Video DAC.

• Video DAC: eight-bit Digital to Analog Converter. Converts the video digital data to analog RGB.

• Master Microprocessor (68070 MPU): central control microprocessor. Manages all functions and data of the MMC Unit.

• Slave Microprocessor: control of port 1, RC5 decoding, and attenuation control.

• System ROM (512k bytes): stores CD-RTOS software executed by Master Microprocessor (68070).

• NV RAM/Clock: Random Access Memory containing data for system configuration at system boot.

The Main Microprocessor (68070) manages all of the MMC Unit's activity and data. All the data transfer (control, video, and audio) within the MMC board is over the 16-bit data bus and 24 bit-address bus structure. The system

ROM's (4 in the CDI601/602 professional CDI players and one in the CDI910) contain the operating system (CD-RTOS), the user shell and the service shell.

The NV RAM contains the Configuration Status Description (CSD) and settings of the player shell. The CSD allows an application to determine what devices are available and contains entries for each available device. When the player is turned On, and after reset, the kernel (operating system house-keeping routine) stored in ROM is always executed by the microprocessor.

Audio Processing

The audio processing path is illustrated in the simplified Audio Processing Block diagram, *Figure 26-10*. The HF (high-frequency information) is read from the disc, decoded (demodulated) in the Decoder (SAA73 IO) and transformed by the ADOC chip (PCF3523) into a serial data stream as DOBM (Digital Output Bi-phase Mark Code).

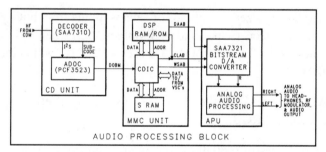

Figure 26-10. The audio processing path is illustrated in this simplified Audio Processing Block Diagram.

The DOBM signal is sent to the CD interface circuit (CDIC) on the MMC panel. The heart of the CD interface consists of the CDIC (IMS66490) and the DSP (Digital Signal Processor) IC's. The CDIC in conjunction with the DSP determines the type of DOBM data received.

If the data is CD-DA the CDIC is switched to the transparent mode. That is, the DOBM is converted back to the I2S format, but there is no other decoding or data management of CD-DA. The CD-DA signals are sent to the APU for digital to analog conversion.

The video data and ADPCM audio data are routed under the control of the main Microprocessor data bus. The ADPCM audio can thus be memory managed to allow synchronization with the video information. The ADPCM audio data is decoded using both the CDIC and the DSP circuits. The decoded ADPCM or PCM (CD-DA) is applied to the D/A converter in accordance with the I2S format.

The digital information is now converted into an analog audio signal by the Bitstream D/A (Digital to Analog) Converter. The APU also provides additional analog audio processing, such as volume control and mixing for CDI applications. The L (Left) and R (Right) audio signals are applied to the Headphones circuit, RF Modulator, and Analog Audio Output jacks (for stereo amplifier or TV monitor).

Video Signal Processing

The Video Processing Block diagram (*Figure 26-11*) shows the overall signal flow for developing the video signal. The HF information coming off the disc is processed the same way as in the Audio Processing circuit. The difference in the processing takes place on the MMC Unit in the CDIC. Data (16 bits), under the control of the Main Microprocessor, is sent to the Master and Slave VSC (Video and System Controller) circuits to develop the a and b video layers to be displayed. The VSC's output both video planes in digital form (8 bits) to the VSD (Video Synthesizer). The VSD receives the encoded image data at a rate of 7.5MB per second from the two VSC'S.

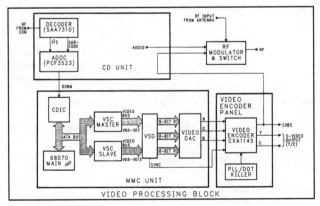

Figure 26-11. *This illustration of the Video Processing Block shows the overall signal flow for developing the video signal.*

The VSD decodes the RGB, CLUT, or DYUV encoded data and adds blanking, weighting, and visual effects (dissolves, wipes, and mosaic transitions) to the data.

The cursor and backdrop are also developed and added to the decoded video in the VSD. The decoded video data is then passed to the Video DAC as eight-bit parallel data for each component (Red, Green, and Blue). The DAC converts digital RGB to analog RGB.

The analog RGB and sync signals are transferred to the Encoder Panel where RGB is converted to composite video (CVBS) and S-Video (Y/C) signals. The composite video and analog audio are also modulated to provide RF (channel 3 or 4) to a standard TV receiver. The Dot Killer circuit is used to remove dots (due to chroma phase errors present with non-interlaced video) in the picture when non-interlaced signals are generated by the decoding circuits.

Part 3, the final part of this series of three articles on CD-I, which will appear in a future issue, will cover troubleshooting and diagnostics built in to the CDI910.

Glossary

ADOC: Audio Digital Output Circuit

ADPCM: Audio Digital Pulse Code Modulation

APU: Audio Processing Unit

CD-DA: Compact Disc Digital Audio

CDI: Compact Disc Interface

CDIC: CD Interface Circuit

CDM: Compact Disc Motor

CD-RTOS: Compact Disc Real Time Operating System

CLUT: A method of encoding CD data

CSD: Configuration Status Description

CVBS: Composite Video

DAC: Digital to Analog Converter

DOBM: Digital output Bi-Phase Mark Code

DSP: Digital Signal Processor

DYUV: A method of encoding CD Data

EFM: Eight to Fourteen Modulation

FTS: Favorite Track Selection

HF: High Frequency

IIS: Inter-IC Signal Bus. The following lines make up this bus:

CLAB: Clock Line

DAAB: Data Line

I2S: Same as IIS

MMC: Multi Media Controller

OPU: Optical Pickup Unit

Photo-CD: A CD system that provides for storage and retrieval of photographs on CD

RGB: Red, Green, Blue Video Signals

SPI Bus: Serial Peripheral Interface

TOC: Table of Contents. The list of the disc's contents which is digitally encoded on the disc. Includes information on which kind of disc this is so that the system can handle it properly.

User Shell: Information that is displayed on the screen of a monitor or TV that provides the user with information about the software that is operating at the moment, and allows the user to select some function of that software.

VSC: Video and System Controller

VSD: Video Synthesizer

Chapter 27

Compact Disc Interactive—Part 3

By Marcel R. Rialland

Serviceability is an important aspect of the development of any product. This is especially important in the case of a high tech product such as Compact Disc Interactive. The Philips CDI91O player includes some features specifically designed with service in mind.

To begin with, the CD-I player is based on the Compact Disc. Therefore, the basic skills for troubleshooting and repairing a compact disc player may be applied to the CD-I player. For example, the CD Servo system must initiate a start-up procedure. The disc can not be read and decoded if there is a failure in the start-up mode. Also, a problem in either the focus or radial servo circuits can cause an error in reading a disc. The Service Shell is very useful in finding problems in the start-up and servo circuits.

The Service Shell

CD-I players have a diagnostic software system built in that allows the technician to perform certain tests with no external test equipment. This software system, along with the information on the screen that allows the technician to gain access to it is called the service shell.

If the player's operating system is working, the software will allow implementation of the service shell. The service shell performs tests of the player's compact disc motor (CDM), servo, video, and audio circuits.

The service shell is implemented by turning power on after placing a jumper across pins 2 and 3 of Port 1 (mouse port). The test mode can also be implemented by using the Philips Service Shell Jumper Plug, part number 4835 310 57148. Only a compact disc digital audio (CD-DA) disc should be used when testing the CDM, servo and audio circuitry in the service shell. No disc is needed for the video test.

The screen shown in *Figure 27-1* is displayed when the service shell is started. From this shell all of the service shell tests can be performed, as long as an icon is highlighted. The cursor, which is in the shape of a wrench, is used to make service selections via the remote control.

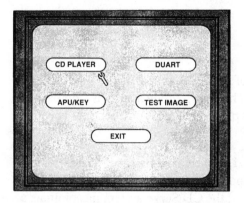

Figure 27-1. The screen is displayed when the service shell is started. From this shell all of the service shell tests can be performed, as long as an icon is highlighted. The cursor, which is in the shape of a wrench, is used to make service selections via the remote control.

Activating the Test Image icon displays color bars to test the video circuits. Selecting the CD Player Icon opens the sub menu shown in *Figure 27-2*. When this menu is selected, the communication channel with the Drive Microprocessor is checked. A message is displayed giving the result of this check (either OK or NO RESPONSE).

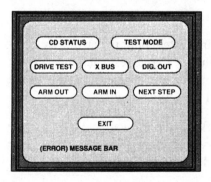

Figure 27-2. Selecting the CD Player Icon opens this sub menu. When this menu is selected, the communiation channel with the Drive Microprocessor is checked. a message is displayed giving the result of this check (either OK or NO response.

In addition to the servo test modes, such as those found on CD players, this menu includes test modes for the communication buses. An error message is displayed if there is a failure in any of the tests. One of the keys around the joystick of the remote transmitter must be keyed to remove the error message. In some cases, further tests are disabled if an error occurs. For example, if the focus test fails, the turntable servo test can not be started. This menu allows all of the Start-up and CDM functions to be tested. The CDM should also be observed during these tests.

The X bus test checks the communication channel between the compact disc interface circuit (CDIC) and the CD Unit's Drive Microprocessor. The DIG OUT test checks whether or not the CDIC receives a Digital Output signal. The most extensive test in this menu is the CD DRIVE TEST. A disc (CD-DA) is needed for this test. TheX BUS and DIG OUT tests are not active when the CD Drive Test is initiated. Selecting the EXIT button returns the player to the CD menu.

The DRIVE TEST consists of the following steps:

• Mode 0: The software release number of the Drive Microprocessor is displayed in the box at the top left of the screen (CD STATUS button). Mode 0 is displayed in the box at the top right of the screen (Mode button). During the CD DRIVE TEST, this icon displays the current Mode. In Mode 0, the ARM IN and ARM OUT tests are active. Selecting NEXT STEP initiates Mode 1.

• Mode 1: In Mode 1 the Drive Microprocessor performs the focus start-up. If focus is achieved (a disc must be present), the message IN FOCUS appears in the status button. Otherwise, the message NO FOCUS appears after 16 focus attempts. In that case (no focus found), the test returns to Mode 0. When focus is achieved, selecting NEXT STEP initiates Mode 2.

• Mode 2: The turntable motor rotates and is controlled by the rough HF (high frequency) (turntable motor servo lock). Moving the CDM arm (by hand) outwardly slows the disc down. If an error occurs, the test returns to Mode 0. Selecting NEXT STEP in Mode 2 brings the player to Mode 3.

• Mode 3: Mode 3 allows the control of the radial arm. If the radial arm servo is operating, you can select ARM IN and ARM OUT to radially move the CDM arm toward the inside or outside of the disc in small jumps. If an error occurs, the test returns to Mode 0. NEXT STEP in Mode 3 puts the player in the normal playing Mode (the test jumper must be removed).

The APU/Key Test

Another sub-menu started from the Main Menu is the APU/KEY TEST. The screen appears as shown in *Figure 27-3* when the APU/Key test is implemented. This is a combined menu. The attenuation can be changed via this menu and the remote control and player keys can be tested. There are three buttons for every attenuation path on the screen. Two of them can be selected (to increment/ decrement) and one is used to display the current attenuation value for the path. There is also a MONO/STEREO button on the screen. In STEREO, two attenuation paths are disabled (left to right and right to left). In MONO all attenuation paths are enabled. Maximum attenuation is reached at value 47.

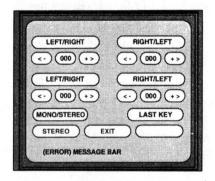

Figure 27-3. This is the way the screen appears when the APU/Key test is implemented. This is a combined menu. The attentuation can be changed via this menu and the remote control and player keys can be tested.

A CD audio disc is needed for the attenuation test. The test routine starts playing the disc when entered. The Key test is used to check the Remote Control and Front Panel Keys. When a key is pressed, text appears on the Key Button on the right side of the screen, identifying the button pressed. The text disappears when the key is released.

Low Level Test

Another diagnostic tool included in the design of the CDI910 player is the Low Level (LL) Test. This test is used to check major functions of the multi media controller (MMC) board. The MMC board is replaced as a module when defective. The LL Test should be performed if the player shell or service can not be accessed. If the Low Level MMC test indicates a fault, the MMC Unit should be replaced. If the Low Level Test cannot be initiated, check the power sources to the MMC board. If all supplies are present, replace the MMC Unit.

The LL Test is implemented in the boot software of CD-RTOS. It does not need a lot of hardware to run. The test can be performed using a VT-100 terminal or the Low Level Test PCB (Philips part number ST1479). A Personal Computer with a VT-100 terminal emulation program may be used as a VT-100 terminal. The LL Test displays the header and release number and checks the VSC (Video and System Controller), ROM, NVRAM, DRAM, CDIC, and Slave Microprocessor (68HC05).

CDI910 Service

All circuit board assemblies can be serviced to the component level except
the MMC board. Due to the complexity of the MMC board, replacement is
recommended when defective. The CDM-9 CD Mechanism is also replaced
as an assembly. Disassembly instructions and exploded views are provided in
the service manual.

Audio Section Troubleshooting

A problem in the CD-I player can be isolated to a particular circuit by
carefully observing the symptoms. For example, if there is no audio, but a
picture from a CD-I disc is displayed, it is obvious that the CD servo and
decoding circuits are functional. Therefore the fault can be isolated to the
audio decoding circuits only.

Or the symptom may be just the opposite: the audio circuits may be working,
but not the video. Again, the servo circuits are functioning. Troubleshooting
of the video decoding should then be followed.

When there is a symptom indicating a fault in the CD Drive circuitry,
troubleshooting techniques used in CD players can be followed since the CD
Unit portion of the CD-I player is basically a CD player, as shown in *Figure 27-4.*

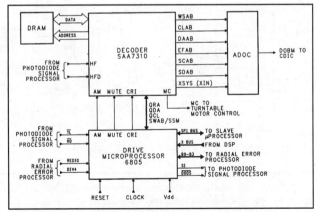

Figure 27-4. *When there is a symptom indicating a fault in the*
CD Drive circuitry, troubleshooting techniques used in CD
players can be followed since the CD Unit portion of the CD-I
player is basically a CD player, as shown here.

If the CDM does not start, check for Vdd, clock, and reset on the CD Drive Microprocessor. If these signals are present, perform the X-bus test in the Service Shell. Also check for activity on the X-bus and SPI-bus. If there is a communication failure, proceed with the MMC Low Level Test to determine if there is a failure in the system control circuitry. If the communication buses are functional, check the CD servo circuits using the Service Shell test modes. If there is a failure in the servo test modes, further checks with a DVM should reveal the fault.

If the servo circuits are functioning, check the decoder circuits. Activity should be seen on the I²S (DAAB, CLAB, and WSAB) and subcode (SCAB and SDAB) lines from the SAA7310 Decoder IC. If there is no activity, check the supply (Vdd) and input signals (HF, XIN). If there is activity on each line, check the ADOC circuitry. When the servo and decoder circuits are functioning properly, there may be a problem in the Audio Processing Unit (see *Figure 27-5*). The APU Panel plugs in to the CD Unit circuit board. Inputs and outputs can be checked at the connectors on both the bottom and the top of the board. Extension connectors (part number 4822 321 22268; requires 2 for service) are available to gain access to the bottom connectors.

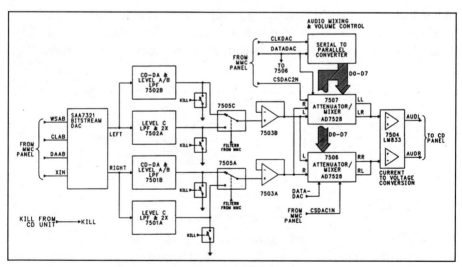

Figure 27-5. *If there is no audio, but the servo and decoder circuits are functioning properly, there may be a problem in the Audio Processing Unit.*

Video Section Troubleshooting

The symptoms displayed can help the technician isolate the problem to a particular circuit. The following examples illustrate how a fault can be isolated (see *Figure 27-6*).

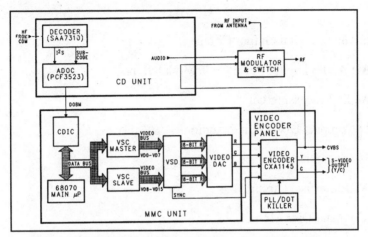

Figure 27-6. *Familiarity with the video processing block diagram shown here may aid a service technician in diagnosing CD-I video problems.*

Player Shell Displayed But No Video

If this condition exists, the Video Encoder Panel and video analog circuits are functioning. Since the player shell is displayed, the video synthesizer circuit is functioning. However, there must be a fault on the MMC Panel, since this is where the video decoding takes place. Thus the problem is isolated to the video decoding section on the MMC Panel.

No Video

This symptom may be caused by either the Video Encoder Panel or by the MMC Panel. A few voltage and signal measurements can quickly isolate the problem. Hint, also check all the video outputs: S-video, RF, and CVBS. If video is not present from any output, check the RGB, composite sync, and power source inputs from the CD Unit to the Video Encoder Panel. If these signals are present, the MMC Panel's decoder circuits are functional.

Final Checks

Make sure to check all player functions after completing a repair on a product such as the CDI player. Also perform any safety checks outlined in the service manual before returning it to the customer.

Glossary

ADOC: Audio Digital Output Circuit

ADPCM: Audio Digital Pulse Code Modulation

APU: Audio Processing Unit

CD-DA: Compact Disc Digital Audio

CDI: Compact Disc Interface

CDIC: CD Interface Circuit

CDM: Compact Disc Motor

CD-RTOS: Compact Disc Real Time Operating-System

CLUT: A method of encoding CD data

CSD: Configuration Status Description

CVBS: Composite Video

DAC: Digital to Analog Converter

DOBM: Digital output Bi-Phase Mark Code

DSP: Digital Signal Processor

DYUV: A method of encoding CD Data

EFM: Eight to Fourteen Modulation

FTS: Favorite Track Selection

HF: High Frequency

IIS: Inter-IC Signal Bus.The following lines make up this bus:

CLAB: Clock Line

DAAB: Data Line

EFAB: Error Flag Line

SCAB: Subcode Clock Line

SDAB: Subcode Data Lines

WSAB: Word Select Line

XSYS: System Clock Line

S2S: Same as IIS

MMC: Multi Media Controller

OPU: Optical Pickup Unit

Photo-CD: A CD system that provides for storage and retrieval of photographs on CD

RGB: Red, Green, Blue Video Signals

SPI Bus: Serial Peripheral Interface

TOC: Table of Contents. The list of the disc's contents which is digitally encoded on the disc. Includes information on which kind of disc this is so that the system can handle it properly.

User Shell: Information that is displayed on the screen of a monitor or TV that provides the user with information about the software that is operating at the moment, and allows the user to select some function of that software.

VSC: Video and System Controller

VSD: Video Synthesizer

Chapter 28

Tricks Enhance
Audio Performance

By John Shepler

We know that sound perception is so subjective that instruments don't always tell the whole story. For instance, one portable radio can sound much richer than another, yet response and distortion tests show that the better specs belong to the worse sounding set. How can that be?

The sound of smaller radios is largely determined by the design of the entire system and mostly by the speaker and enclosure. Much of the richness that is perceived comes from reverberation and resonances through the case. A booming low end makes the sound fuller and it helps the mask mid and high frequency distortion. The sound seems to come deeper from within the loudspeaker.

You can run some simple experiments to prove this, Try hooking various types and sizes of loudspeakers through the earphone jack. Note the change in sound quality. The better speakers, especially without enclosures, often sound worse. Some of the really good acoustic suspension speakers sound terrible because the amplifier doesn't have enough power to drive the speaker without severe distortion.

Portable radio designers use every trick they can think of to improve the perceived quality of their product. A type of stereo effect can be produced by a mono radio by using large and small speakers on opposite sides if the case. This splits the frequency band so that low frequency instruments, such as drums, are primarily reproduced by the larger speaker on one side of the radio. Higher frequency instruments, such as cymbals, are most transmitted by the smaller speaker on the other side. On some songs, the stereo effect can actually sound quite good.

Another simple trick is the "wide" versus "normal" stereo switch. This is easy to implement with a switch that simply reverses the phase to one of the speakers. Normally, stereo speakers are driven in phase so that mono materials drives the cones in the same direction at the same time. This puts singers, commercials, and other mono material squarely between the speakers. Reversing the phase causes the speakers to cancel each other so that the portion of the stereo stage between them disappears. The effect sounds wider because the sound is pushed out to the speaker location rather than mixing in the middle. This just goes to prove that sometimes technically incorrect actually sounds better. *Figure 28-1* shows the hookup you can try with any stereo setup.

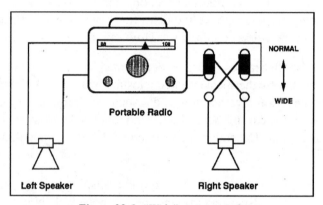

Figure 28-1. *"Wide" stereo switch.*

Car audio systems have their own tricks for sounding great. These systems are designed to enhance the performance of the overall perceived sound rather than optimizing any particular component. Some years ago, I remember being astounded by how horrible $300.00 Ford and GM car radios, especially the AM band, sounded through earphones on the bench. Put them back in the vehicle and they sounded terrific. Those rear deck 6x9 speakers reverberating through the trunk made all the difference.

Home stereo systems offer less opportunity for this type of acoustic manipulation. The receiver is separate from the speakers and is rated on its own performance, notably power output, distortion, frequency response, separation, and RF performance. Speakers are almost all sealed wood boxes with subtle increases in transparency as you go up the range in price.

Most audio manipulation on home setups is done with parametric and graphic equalizers, pointing out how much perfectly flat component frequency response is really worth. Speaker placement does make a big difference. One long-time trick has been to stand the speakers in the corners of the room to gain extra bass response by using the walls as a sounding board.

Another trick that adds richness to the sound is adding electronic reverberation with a separate control unit. The older ones use a spring and acoustic transducer to provide the delay. Newer designs use digital techniques such as BBD (bucket brigade devices) or DSP (digital signal processing). A small amount of reverb gives that rich sounding concert hall effect heard on live albums.

Superb sound to the consumer isn't always the same as perfect technical specifications. Sometimes the ear can be tricked.

Chapter 29

Magnetic Recording Principles: Audio and Video

By Lamar Ritchie

Audio and video tape recording have provided immense opportunities for entertainment in the home. They have also provided innumerable opportunities and problems for service technicians.

An understanding of magnetic recording tape construction and recording principles can help a service technician in diagnosing a problem found in audio or video recorders.

The Principles of Magnetic Recording

The basic principle of all magnetic recording is the same, whether the information recorded is video or audio. A thin plastic "tape," coated with very fine magnetic particles such as iron oxide or chromium dioxide, is moved past an electromagnetic "head" at a constant velocity.

During recording, a variable current is sent to the head, producing a variable intensity magnetic field which in turn produces regions of varying degrees of alignment of the magnetic domains on the surface of the tape. During playback, movement of the recorded tape across the head gap causes a varying ac voltage to be induced in the head's coil.

Audio Recording Requires a Bias Signal

The audio ac voltage cannot simply be applied to the magnetic head as is. There are two reasons for this:

• the tape's surface, being a ferromagnetic material does not have linear characteristics, and

• the head produces an output whose amplitude is not linear with frequency.

The biggest problem caused by the nonlinear characteristic of the tape's ferromagnetic material occurs at the lower end of the characteristic curve, where alignment of the magnetic domains does not start to occur until some definite, non-zero amount of magnetizing field is applied (*Figure 29-1*). To place the audio signal in the linear region, an ultrasonic bias, in the range of 60KHz to 100KHZ, is used (*Figure 29-2*).

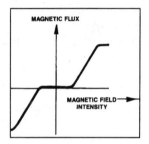

Figure 29-1. The biggest problem caused by the nonlinear characteristic of the tape's ferromagnetic material occurs at the lower end of the characteristic curve, shown here, where alignment of the magnetic domains does not start to occur until some definite, non-zero amount of magnetizing field is applied.

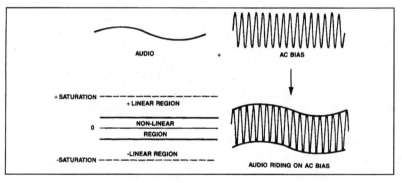

Figure 29-2. To place the audio signal in the linear region, an ultrasonic bias, in the range of 60KHz to 100KHz, is used.

The audio signal is superimposed on the high-frequency ac bias, producing a variation of the ac signal, the peaks of which do not extend into the non-linear region of the tape's characteristics. The actual signal that is applied to the head has much less variation in it than shown in the diagram. For purposes of the diagram, the variation was magnified for clarity. The actual level of the bias ac may be several volts and the audio in millivolts. The ac bias actually drives the head into magnetic saturation.

In fact, the bias is not recorded, or recorded very little. The bias drives a particular magnetic domain of the tape around the hysteresis loop several times and as the tape moves on and the magnetic field applied to it falls to zero, it comes to rest at a certain magnetization depending on the signal current.

The Magnetic Recording Head

Figure 29-3 shows how the head places the varying alignment of the magnetic domains on the tape. The head contains a small non-magnetic gap that is placed in contact with the tape as it goes by. The two ends of the gap act as the poles of the electromagnet. The tape has a relatively high magnetic permeability and acts as a low reluctance path for the lines of force across the gap. As the lines of force vary in intensity, the degree of alignment of the tape magnetic domains is varied.

PROFESSIONAL GRADE EQUIPMENT

REEL TO REEL RECORDERS

TAPE WIDTH	TRACKS	SPEED (in./sec.)	SPEED (cm/sec.)
2" (5.08cm)	24-48	15,30	38.1/76.2
1" (2.54cm)	2-16	15,30	38.1/76.2
1/2" (1.27cm)	2/4/8	15,30	38.1/76.2
1/4" (.64cm)	2	7-1/2,15	19.1/38.1

CONSUMER GRADE EQUIPMENT

CASSETTE TAPE

TAPE WIDTH	TRACKS	SPEED (in./sec.)	SPEED (cm/sec.)
1/4"	1/2/4	1-7/8, 3-3/4	4.8/9.5
		7-1/2,15	19.1/38.1

8-TRACK CARTRIDGE

TAPE WIDTH	TRACKS	SPEED (in./sec.)	SPEED (cm/sec.)
1/4" (.64 cm)	8	3-3/4	9.5

CO-PLANAR HUB CASSETTE

TAPE WIDTH	TRACKS	SPEED (in./sec.)	SPEED (cm/sec.)
.15" (.38 MM)	2/4	1-7/8	4.8 (3-3/4 on some 4-track recorders)

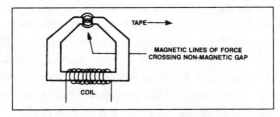

Figure 29-3. *The head places the varying alignment of the magnetic domains on the tape.*

The actual size of these small magnetized areas of the tape is called the "recorded wavelength." The recorded wavelength is determined by the *tape speed and signal frequency* (*Figure 29-4*).

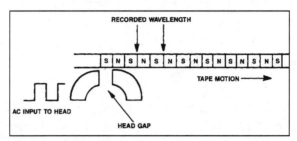

Figure 29-4. The actual size of the small magnetized areas of the recorded tape is called the "recorded wavelength." The recorded wavelength is determined by the tape speed and signal frequency.

The Head Nonlinearity Problem

One of the basics that most electronics courses teach is that the induced voltage in a conductor by generator action is proportional to the change in magnetic field intensity in the vicinity of the conductor. This is, in turn, proportional to the speed of motion of the conductor through a magnetic field. This principle also applies to the head during playback of the signal.

Higher frequency signals cause a faster change in the field for the same tape speed, and therefore, a higher output voltage from the head. As the frequency increases, the output of the head will increase by 6dB per octave. The output, then, must be equalized by a filter having opposite characteristics.

As for the range of frequencies that can be recorded on magnetic tape, the upper limit is determined by the speed of the tape motion relative to the head (writing speed), and the width of the head gap. Maximum output will occur when the recorded wavelength is twice the width of the head gap. From there on, the output will decrease and reach minimum when the recorded wavelength is equal to the width of the head gap. At this frequency, both a north and a south pole are positioned within the gap and the fields will thus cancel. *Figure 29-5* illustrates the change in relative output of the head as the frequency changes.

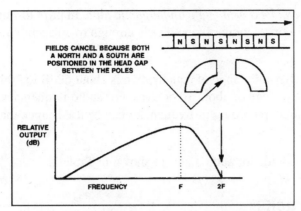

Figure 29-5. The illustration shows the change in relative output of the head as the frequency changes. The recorded wavelength is equal to the writing speed divided by the signal frequency.

The recorded wavelength of a signal is equal to the writing speed divided by the signal frequency.

Limitations in the Recording Process

Figure 29-5 indicates the response under ideal conditions and other factors have influence. Since the slope rises 6dB/octave it is important to have very little noise to extend the usable frequencies. The tape itself is the biggest factor here.

Random nonuniformities in the oxide coating cause random noise. On poor tape, oxide not bound well will come off and worsen the problem. Other factors that limit performance are:

• *"Fringing" of the field.* A smearing of the recorded pattern because the field extends a little outside of the tape.

• *Self erasure of higher frequencies.* As the domains swing around quickly, inertia carries them past the correct alignment and they come to rest at a more random point (ac fields are used to erase tapes for this reason).

• *Separation losses resulting from imperfect contact between the tape and the head.* This loss amounts to about 55dB for a tape-to-tape separation of one recorded wavelength.

• *Printthrough can be a problem with magnetic tape.* In print through, the magnetic pattern on a length of tape bleeds through to adjacent wraps of the tape.

At present, the dynamic range of equalization is about 60dB to 70dB. This gives a frequency range of about 10 octaves. For audio frequencies, 20Hz to 20KHz, then, direct recording of frequencies can be used, since that represents 10 octaves.

Some of the standards for audio tape are shown in Table 1.

Multiple Tracks

To obtain maximum recording time from a given length of magnetic tape, most consumer audio recorders use multitrack recordings. Of course, for stereo sound, two channels (tracks) are required for each recording. The tracks for each recording usually take up only half of the width of the tape so that the tape can be turned over to record on the "back side."

To help prevent crosstalk between channels, the tracks are separated for some formats. The track formats for reel-to-reel and the now obsolete 8-track recorders are as shown in *Figure 29-6.*

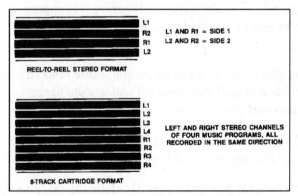

Figure 29-6. The track formats for R-R and the now obsolete 8-Track recorders.

For 1/2 track (monophonic) reel-to-reel, the top half of the tape width is simply side 1 (track 1) and the bottom half of the tape is side 2.

The mono reel-to-reel is not compatible with the stereo format, because front and rear channels will overlap. The separation between tracks is not needed much, today, because of better heads and tape. The cassette format, therefore, has the tracks for each channel adjacent to each other. This arrangement provides compatibility, meaning a mono player can play a stereo recording, and both channels will play through the one amp.

Improvements in Fabrication of the Tape

As for the tape itself, old tape was dull on the oxide side with a shiny backing. Newer tape may be textured, or "roughed up" on the non-oxide side to reduce slippage, and may be highly polished on the oxide side to get a smoother surface that will reduce head wear. The tape will be constructed something like that shown in *Figure 29-7*.

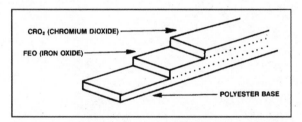

CRO₂ (CHROMIUM DIOXIDE)

FEO (IRON OXIDE)

POLYESTER BASE

Figure 29-7. Audio tape is constructed something like this.

Every time a tape is recorded in a tape recorder, it is first erased. To cause this erasure, a high level of the ac bias signal is applied to an 'erase' head at the beginning of the tape path. Lower frequency ac (60Hz) can be used over a large length (entire tape) in a "bulk eraser." The tape is either placed on the eraser for a short time, or a handheld bulk eraser is moved around the reel or case a few times.

The same heads are generally used for playback and recording. Some machines, however, have separate play and record heads, primarily to make immediate monitoring of recorded sounds possible.

Video Tape Recording

Articles in future issues will discuss video tape recording principles.

Chapter 30

Digital Tuners Have Arrived
By John Shepler

The age of the digital tuner has arrived. While over-the-air digital audio broadcasts (DAB) are still years away, CD-quality audio is becoming available now via cable television. The digital tuner is the last piece needed to complete an all-digital home audio system.

Radio on cable has been available for many years. The cable operators provide a spectrum of FM services consisting of radio stations and satellite-delivered audio. The appeal of the satellite services is that they have no commercials and often no announcements.

The limitation to standard cable radio is that the signal has to be delivered through an RF carrier with the standard stereo modulation first established nearly 30 years ago. This is true even if the source material is Compact Disc.

The new digital audio services dispense with standard radio transmission completely. A digital program is generated at a remote studio site and uplinked as a digital bitstream to a satellite transponder. At the receiving end, the cable operator extracts the signal and relocates it to a carrier frequency compatible with the cable signal assignments. In the home, the bitstream is demodulated from the cable carrier, kept in digital format, and finally converted to analog audio by a digital to audio converter in the tuner or power amplifier.

The obvious advantage to digital audio is that perfect quality can be maintained from the digitized source material in the studio to the D/A converter in the home. The digital bitstream can also be recorded on to digital audio tape (DAT) or digital compact cassette (DCC) before converting to analog.

Digital's less obvious advantage is that additional data can be sent along with audio bitstream. For instance, a remote control unit for the tuner can display information about the music on a liquid crystal display. In the future, a listener might program the system to watch for a particular selection or connect and record it in the middle of the night.

There are two major services now expanding nationally on cable systems. They are Digitial Music Express (DMX) with 30 channels and Digital Cable Radio (DCR) with 19 channels, soon to be 29, with 250 planned in 10 years. These services offer a wide variety of specialized music formats and stereo simulcasts of premium TV services, such as HBO. DCR has also experimented with a pay-per-listen service for a special concert event.

Figure 30-1 shows how a digital home audio system might be configured. While the delivery service now is through cable TV hookup, fiber optic transmission is in the future for both audio and video. Satellite and over-the-air digital audio broadcasting standards are in the process of being standardized and should be available in the home in a few years.

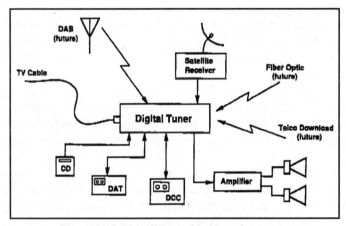

Figure 30-1. *The All-Digital home audio system.*

Another interesting concept has been developed by Bellcore Labs for the telephone companies. That's right, it is feasible to send digital audio at 1.544 megabits per second over most phone lines from the central office to the home. Excluded are 25% of the lines that have inductive loading coils to neutralize high capacitance. The digital audio system called ADSL or asymmetrical Digital Subscriber Line also includes a reverse channel so that the consumer can "dial up" the desired audio service or send other information back to the supplier. While still experimental, this service might become available in as little as two or three years. All-digital audio is a technology whose time has come... Look for integrated digital audio components, such as tuners and receivers with digital I/O, to become available soon.

INDEX

C

T

ES&T Presents
TV Troubleshooting
& Repair

ES&T Presents TV Troubleshooting & Repair presents information that will make it possible for technicians and electronics hobbyists to service TVs faster, more efficiently, and more economically, thus making it more likely that customers will choose not to discard their faulty products, but to have them restored to service by a trained, competent professional.

Originally published in *Electronic Servicing & Technology*, the chapters in this book are articles written by professional technicians, most of whom service TV sets every day. These chapters provide general descriptions of television circuit operation, detailed service procedures, and diagnostic hints.

ES&T Presents
Computer Troubleshooting
& Repair

ES&T is the nation's most popular magazine for professionals who service consumer electronics equipment. PROMPT® Publications is combining its publishing expertise with the experience and knowledge of *ES&T's* best writers to produce a new line of troubleshooting and repair books for the electronics market.

Compiled from articles and prefaced by the editor in chief, Nils Conrad Persson, these books provide valuable, hands-on information for anyone interested in electronics and product repair. *Computer Troubleshooting & Repair* is the second book in the series and features information on repairing Macintosh computers, a CD-ROM primer, and a color monitor. Also included are hard drive troubleshooting and repair tips, computer diagnostic software, preventative maintenance for computers, upgrading, and much more.

Troubleshooting & Repair
226 pages • paperback • 6 x 9"
ISBN: 0-7906-1086-8 • Sams: 61086
$24.95

Troubleshooting & Repair
256 pages • paperback • 6 x 9"
ISBN: 0-7906-1087-6 • Sams: 61087
$24.95

Speakers For Your Home and Automobile, *2nd Edition*
Gordon McComb, Alvis & Eric Evans

The cleanest CD sound or the clearest FM signal are useless without a fine speaker system. *Speakers for Your Home and Automobile* will show you the hows and whys of building quality speaker systems for home or auto installation. With easy-to-understand instructions and clearly-illustrated examples, this book is a must-have for anyone interested in improving their sound systems.

The comprehensive coverage includes:

Speakers, Enclosures, and Finishing Touches
Construction Techniques
Wiring Speakers
Automotive Sound Systems and Installation
Home Theater Applications

Audio
192 pages • paperback • 6 x 9"
ISBN: 0-7906-1119-8 • Sams: 61119
$24.95

Complete Guide to Audio
John J. Adams

Complete Guide to Audio was written for the consumer interested in sound systems. With comprehensive, simple explanations, it answers questions you may have asked salespeople but were unable to get answers for.

In addition, this book explains common problems you may experience while setting up your home entertainment center. This book will help you make successful purchasing decisions and demystify the jungle of wires and connections in your audio system.

Topics include: Audio Basics, Sound, Stereo-phonics, Home Theater, Amplifiers and Preamplifiers, Receivers and Surround-Sound, Cassette and CD Decks, DVD, MiniDisc and Phonographs, Speakers, Headphones and Microphones, Computer Sound, Brands and Choices, Hookups and Accessories.

Audio
163 pages • paperback • 7-1/4 x 9-3/8"
ISBN: 0-7906-1128-7 • Sams: 61128
$29.95

CALL 1-800-428-7267 TODAY FOR THE NAME OF YOUR NEAREST PROMPT PUBLICATIONS DISTRIBUTOR

Prices subject to change.